GEOLOGY OF WISCONSIN AND UPPER MICHIGAN

GEOLOGY OF WISCONSIN AND UPPER MICHIGAN

INCLUDING PARTS OF ADJACENT STATES

RACHEL KREBS PAULL
Alverno College

RICHARD A. PAULL
University of Wisconsin—Milwaukee

KENDALL/HUNT PUBLISHING COMPANY
2460 Kerper Boulevard, Dubuque, Iowa 52001

Library of Congress Catalog Card Number: 76—27036

ISBN 0—8403—1596—1

Printed in the United States of America

To our parents, Lester J. and Ruth Sells Krebs and Ethra H. and Martha Schaller Paull, who encouraged us to learn; and our teachers, especially those who introduced us to Wisconsin geology: Professors Lewis M. Cline, Robert M. Gates, Sheldon Judson, Lowell R. Laudon, Katherine G. Nelson, Frederick T. Thwaites, and Stanley A. Tyler; and our children Kay, Lynn, and Judy who carried on without us while we worked.

CONTENTS

"To a person uninstructed in natural history, his country or seaside stroll is a walk through a gallery filled with wonderful works of art, nine-tenths of which have their faces turned to the wall." *Thomas Henry Huxley*

PREFACE

This book provides an account of the geologic history, and a description of the four major physical provinces of our region. We have assumed that the reader has little or no prior geologic background, but has enough interest in the earth to want to learn. To accomplish this, the book is divided into eight chapters.

The first provides a summary of the basic geologic principles necessary to understand the following sections. The next chapter traces the geologic history of the area from the earliest to the latest event which enables us to provide an interpretation of the present landscape in the following sections. We do this by dividing the region into four physical provinces with distinctive landscapes that reflect the geologic history of that area. For each province we discuss the general characteristics, bedrock geology, glaciation, drainage, and special features worth visiting. The next section provides information on rock, mineral, and fossil collecting, and the final section describes the geologic resources of the region.

Although this is a "beginners" book, we think you will end up with a sophisticated appreciation of the geology of this region. You will find that the landscape is extremely varied and interesting and that your knowledge of the earth and earth processes will bring new pleasures when you travel.

To make the most of this new knowledge we hope you will take to the field and follow the geologic guides to this region that are provided in a companion volume. Get outside, make your observations, and weave them into geologic history and the landscape around you. When you do this, we think you too will discover that geology is fun! Happy geology!

Acknowledgments

When we started this project we thought it would be fun and easy. It was fun but it certainly wasn't easy. We soon found out how little we knew about this fascinating area and how much others knew. We learned from them and we appreciate the education. We are indebted for the work of numerous geologists who published under the auspices of the: U.S. Geological Survey, Illinois State Geological Survey, Iowa Geological Survey, Minnesota Geological Survey, Michigan Geological Survey Division, and Wisconsin Geological and Natural History Survey. Of all the published materials we studied, we found *Geology of Minnesota* edited by P.K. Sims and G.B. Morey (1972), *Geology of Michigan* by J.A. Dorr and D.F. Eschman (1970) and *The Physical Geography of Wisconsin* by Lawrence Martin (1932) to be invaluable and we urge students of this region to acquire them.

One aspect of our studies involved a renewed appreciation of the early geologic workers in this area. We who follow merely turn the rich soil that was broken out of virgin wilderness by true geologic pioneers. There were Henry R. Schoolcraft, David Dale Owen, Increase A. Lapham, James Hall, Thomas C. Chamberlain, Charles R. Van Hise, C.K. Leith, W.C. Alden and Frederick T. Thwaites, to name a few. Lawrence Martin, an early synthesizer, also belongs with the pioneers. We, who follow with Cutter's insect repellent, radiometric age dating, electric typewriters, and precision mapping via satellites, can move across the land at more than 55 miles per hour with restaurants, motels, and campgrounds around every bend in the road. Yet, with all of these advantages, we still can't match the accomplishments of the early workers.

We are also indebted to many individual geologists who generously provided information based on their personal research. These include: Meredith E. Ostrom (Wisconsin Geological and Natural History Survey), Katherine G. Nelson (University of Wisconsin-Milwaukee), Gregory Mursky (University of Wisconsin-Milwaukee), Robert H. Dott, Jr. (University of Wisconsin-Madison), Robert F. Black (University of Connecticut), Gene L. LaBerge (University of Wisconsin-Oshkosh), Thomas S. Laudon (University of Wisconsin-Oshkosh), Carl E. Dutton (U.S. Geological Survey), Paul E. Myers (University of Wisconsin-Eau Claire), Ralph W. Marsden (University of Minnesota-Duluth), Paul K. Simms (U.S. Geological Survey), Paul C. Tychsen (University of Wisconsin-Superior), Timothy J. Rohrbacher (White Pine Copper Company), and Bruce E. Brown (University of Wisconsin-Milwaukee) who also critically reviewed the manuscript. With all this help it is apparent that very little in this volume represents our original work. Consequently we gladly credit our sources for what is right while we accept responsibility for any errors.

We would like to thank the Wisconsin Department of Natural Resources, the Michigan Conservation Department, the Milwaukee Journal, Cave of the Mounds, and the Wisconsin Dells Chamber of Commerce for selected photographs displaying the beauty of this region. We are especially indebted to Mr. Galen Parker of the Wisconsin Department of Natural Resources for assistance with the selection of appropriate photographs. Mr. Eugene M. Roark, Director of Vacation and Travel Services of the Wisconsin Department of Natural Resources, also provided advice.

We appreciate the indulgence and encouragement of all our colleagues at the University of Wisconsin-Milwaukee while we continued this work. We especially wish to acknowledge: the skillful drafting of Mr. Donald Temple, Ms. Donna Arndt, and Mr. Larry Saeger of the Cartographic Laboratory of the Department of Geography, the photographic assistance of Mr. Frank J. Charnon of the Department of Geological Sciences, and the typing and retyping of Ms. Mary H. Polzin, Ms. Shirley Halloran, Ms. Rose Ann Magidson, and Ms. Mary Sonsalla of the Department of Geological Sciences, and the editing of Ms. Joyce C. Grueter and Mrs. Carol G. Stolz, also of the Department of Geological Sciences. Without their help we would still be struggling with a draft manuscript.

We cheerfully acknowledge the hospitality of innkeepers at many crossroads. Their products, when mixed with Wisconsin cheese right from the factory, or pasties from the old mining towns of southwestern Wisconsin and northern Michigan, provided many a pleasant interlude on long geologic days.

We probably should also credit our vintage Volkswagen and Chevrolet for about 25,000 miles of essentially trouble-free travel as we pursued our field studies of Wisconsin and parts of adjacent states through some of the worst and "most unseasonable" weather this region has ever known. However, that's the way it often is when geologists take to the field.

Chapter 1

SOME GEOLOGIC PRINCIPLES

Appreciation of the landscape of any area requires an awareness of the geologic history that created it. This is especially important if you wish to enjoy the beautiful and varied countryside of Wisconsin and the adjacent states. The landscape of this area results from a long and complex history that can be interpreted through the record of past events in rocks and landforms.

To read earth history requires the understanding of geologic principles and the desire to thoughtfully observe the geologic record as it is preserved for us to see. Knowledge of geologic principles is the first step toward a new and fuller appreciation of the earth.

THE ROCK CYCLE

Many of the principles of geology are embodied in the concepts of the *rock cycle* (Fig. 1-1). Conceptually, the rock cycle, like the precrustal history of the earth, starts with magma. *Magma* is a molten liquid composed of ions that forms deep in the crust or upper mantle in response to heat produced by radioactivity. *Igneous rocks* result from the cooling and crystallization of magma, within the earth or at its surface.

Magma may cool within the earth, and as it cools, various minerals crystallize from the melt. The size and type of crystals that form depend upon the composition of the magma and the rate of cooling. A slow rate of cooling will produce coarse-grained rocks, with readily visible crystals. If the original magma has suitable ions present, the first minerals to crystallize will be those that form at higher temperatures. These minerals are rich in iron, calcium, and magnesium but deficient in silica. Gabbro is a rock that contains such minerals. If these early formed minerals sink to the base of the magma chamber, the melt will become progressively depleted in calcium, magnesium, and iron, thereby concentrating the remaining sodium, potassium, and silica. These ions will form minerals that crystallize into granitic rocks.

Magma within the earth can intrude older rocks and crystallize to produce a variety of igneous bodies (Fig. 1-2). These are the plutonic rocks. They may be large masses called *batholiths* or they may be relatively thin and tabular. Tabular igneous rock bodies that are concordant with surrounding rock layers are called *sills,* while those that cut across earlier formed rocks are *dikes.* Another type may be circular plugs or necks that feed volcanoes. In some instances the intrusion of an igneous rock body may severely deform the surrounding rock, while in other cases there will be little deformation.

Magma may also be introduced at the earth's surface as lava flows, volcanic ash, or coarser volcanic ejecta that include bombs and large rock fragments. Magma that reaches the earth's surface cools rapidly and fine-grained rocks result. If the magma is rich in iron and magnesium and deficient in silica, basalt forms. If it is silica-rich, the rock is rhyolite. Some lavas crystallize fast enough to trap gas bubbles. Such void areas

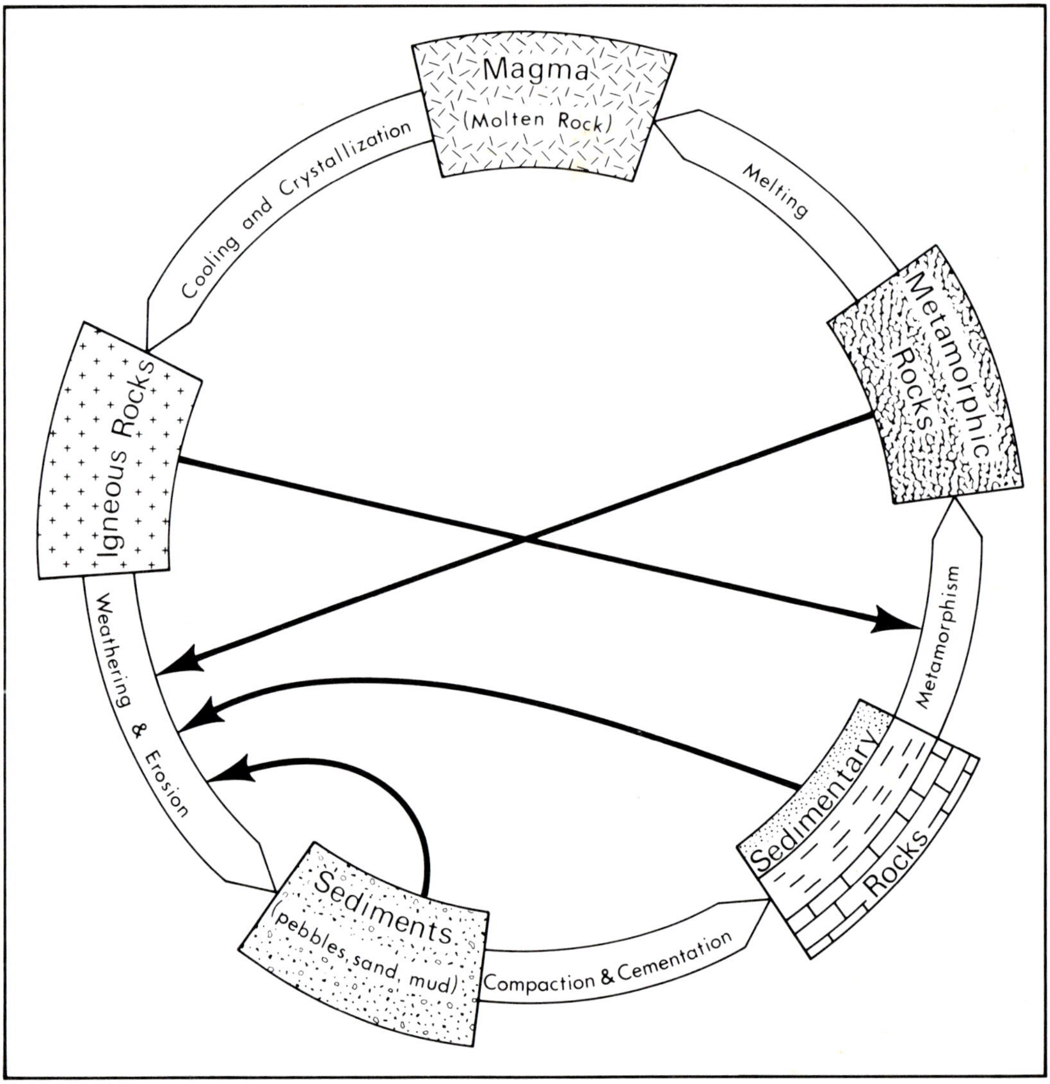

Figure 1-1. The rock cycle. The crossovers shown represent alternatives to the idealized cycle represented by the circular periphery of the diagram.

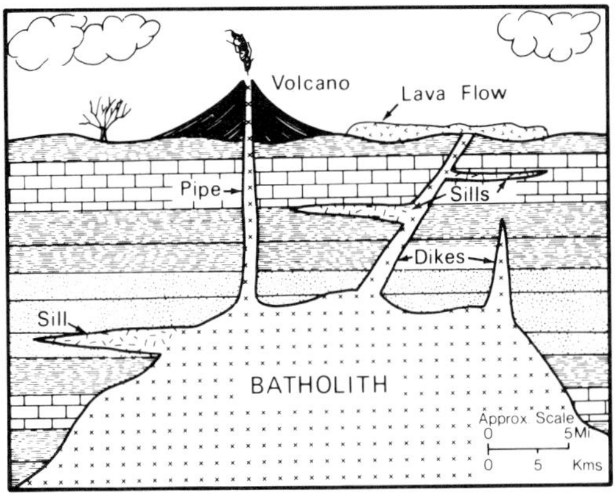

Figure 1-2. Diagrammatic cross section of igneous rock bodies that form within the earth's crust (intrusives) and at the surface (extrusives). The largest intrusive masses are batholiths. Tabular intrusive bodies that are concordant to older rocks are sills; those that are discordant to the surrounding rocks are dikes. Circular, discordant bodies are pipes and these usually feed volcanoes. Material ejected from volcanoes accumulates on the surface as ash falls or coarser deposits. Dikes and pipes that reach the surface provide molten material to form lava flows.

are called *vesicles.* Vesicles that later become filled with minerals are *amygdules,* and igneous rocks with these features are amygdaloidal lavas. Sometimes magma partially crystallizes within the earth and then is extruded to the surface. This results in a fine-grained rock with some large crystals present. Such rocks are called *porphyrys.*

Extrusive igneous rocks on the earth's surface will be subjected to weathering processes immediately after formation. *Intrusive* igneous rocks that form deep within the earth may ultimately be uplifted and exposed at the surface by erosion. Once exposed, they too begin to weather and erode away.

The agents of erosion that are operating today give us insight into how they operated in the geologic past. This is the *principle of geologic uniformitarianism* which can be stated: the present is the key to the past. Knowledge of contemporary erosional processes such as running water, wind, and glaciers is a great aid in interpreting the geologic record.

The erosional agents move earth materials downslope under the influence of gravity to sites of deposition. The lowest places on earth are the oceans, and ultimately all dissolved material, rock fragments, sand, and mud eroded from the surface are deposited in the sea. Sediment accumulates there in great thicknesses and it progressively becomes compacted and cemented into layers of *sedimentary rock.* Pebbles or cobbles make conglomerate, sand forms sandstone and mud becomes mudstone or shale. Similarly, deposits of broken shell debris or mud composed of calcium carbonate (calcite) will form limestone, which may be changed into dolomite by the partial replacement of calcium by magnesium.

Sedimentary rocks, like igneous rocks, may be uplifted and eroded to produce particles that can then be redeposited to form new sedimentary rocks. It is also possible that sedimentary and igneous rocks may be buried deep within the earth and subjected to great heat and pressure. Such conditions will produce changes in the texture and/or mineralogy. Rocks that are modified in this way are *metamorphic rocks.* Metamorphic rocks are also produced by heat associated with large igneous intrusions or by regional structural forces that involve movements of the earth's crust.

Metamorphic rocks generally have a coarser texture and contain minerals that form under higher pressure and/or temperature than those originally present. As a result, metamorphic rocks are usually denser and harder than the original rocks. Often the newly formed minerals are aligned to give the rock a banded appearance, such as we see in gneiss. Mineral alignment may also allow the rocks to part along the plane of alignment, like schist and slate. Once formed, these rocks may also be uplifted and eroded.

During metamorphism, if the temperature continues to increase, rocks will melt to form magma. High temperature laboratory experimentation shows that temperatures over 600 degrees Celsius are required to melt rocks. In the laboratory, as in the earth, rock type and pressure are also important controls on magma formation. Once magma is formed and begins to cool, crystallization forms igneous rocks as the rock cycle continues.

OROGENIC CYCLE

The rock cycle concept implies that rocks are periodically uplifted and eroded to supply sediments which form sedimentary rocks. As the cycle continues, these rocks may be metamorphosed, melted to form magma, and later crystallized as igneous rocks. This aspect of the rock cycle is sometimes referred to as the *orogenic cycle.*

As shown in Figure 1-3, a single orogenic cycle involves a sequence of uplift, erosion, deposition, subsidence, possibly metamorphism, and even melting if subsidence persists to great depths. Melting would be followed by cooling and crystallization and another uplift. As in the rock cycle, the orogenic cycle doesn't always go full circle, for uplifts may interrupt the idealized sequence.

Uplift or subsidence of adjacent parts of the earth's crust are integral to the orogenic cycle. For example, we see that the northern part of Wisconsin and much of northern Michigan was an area of relative uplift during the Paleozoic while areas adjacent to this uplift were subsiding and receiving sediments.

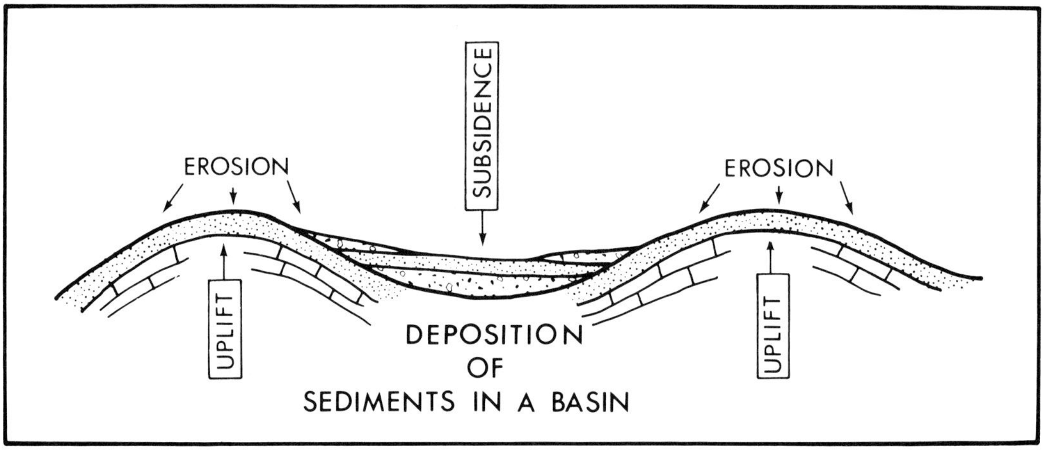

Figure 1-3. The orogenic cycle is illustrated by a cross section of a depositional basin flanked by arches that are eroding and contributing sediments to the basin. As basin subsidence continues, the weight of overlying sediments aids in depressing the crust. At great depth, increasing temperature and the pressure of overlying rocks produces metamorphic changes. Ultimately the deeply buried rock might melt to form magma. At some later time, the original basin might be uplifted to supply sediments to a newly forming depositional basin.

Why some areas of the crust were uplifted while others subsided is often difficult to explain. Furthermore, areas of subsidence may become areas of uplift later on. In some instances the uplifts are adjustments that result when lighter masses of rock, such as sedimentary or granitic rocks, essentially "float" upward relative to associated masses of denser rocks. Uplifts and downwarps also result when compressional forces cause buckling into large-scale upward and downward folds. Another important mechanism that causes adjustments between segments of the earth's crust involves the movement of lighter continental plates of the earth over denser crust and mantle rocks.

All of these mechanisms may have operated in the north-central United States during five billion years of earth history. As you will see later, we know very little about the details of orogenic cycles in this region during the first several billion years, but we believe that many of them were episodes of intensive upheaval that formed major mountain systems. Events of this magnitude are called *orogenies.* Other types of uplift are relatively gentle and they are accomplished without appreciable breakage of the earth's crust.

It is generally accepted that the Lake Superior basin formed, at least in part, in response to the weight of thick outpourings of Precambrian lava flows that were denser than the crustal rocks upon which they accumulated. The lavas ceased to flow before the end of the Precambrian, and the area was uplifted during the Paleozoic. What started the lava flows in this specific area, why did they end, what caused an uplift, and how does the Lake Superior basin relate to other large features that existed on the North American continental plate? These are just a few of the geological questions about this region that lack definite answers, but the search for truth goes on.

As we will describe in more detail later, *arches* (upwarps) and *basins* (downwarps) characterize the Paleozoic history of the Upper Midwest. The arches were slowly subsiding areas between more rapidly foundering basins. The forces that initially created negative areas or basins in the interior of a large continental plate are unknown, but once these basins began to fill with sediment, this weight certainly helped hold them down. However, the entire region was uplifted late in the Paleozoic to initiate the present erosion interval. So we see, our area has a complex and cyclic history of erosion, subsidence, deposition, and uplift that is continuing.

ROCK TERMINOLOGY

The rock cycle provided a convenient way for us to introduce the three major rock types. Each of these types (igneous, sedimentary, and metamorphic) was formed under conditions that imparted textural and compositional characteristics to the rocks that enable us to recognize their mode of formation (Fig. 1-4).

Igneous rocks crystallize from a melt and an interlocking pattern (texture) of minerals results. As previously described, there is a predictable order to crystallization that relates to temperature. At approximately 1,000 to 1,200 degrees Celsius, heavy, dark silicate minerals rich in iron and magnesium form. These are the ferromagnesian minerals, olivine and pyroxene (augite). Also crystallizing at this time is calcium-rich plagioclase feldspar (calcium aluminum silicate). Rocks containing these minerals are called *basic* igneous rocks, and some coarse-grained varieties are peridotite (ferromagnesian-rich), gabbro (ferromagnesians and plagioclase), and anorthosite (plagioclase-rich). Most lava flows in our area are basalt,

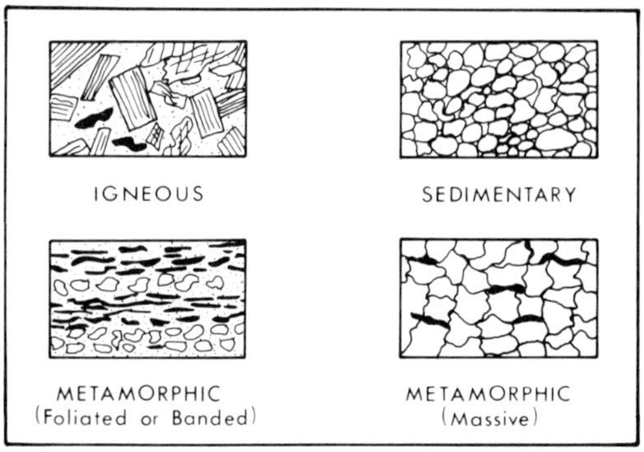

Figure 1-4. Common textures of igneous, sedimentary, and metamorphic rocks. Igneous rocks show an intergrowth of crystals and there are no void spaces between them. Sedimentary rocks are composed of a variety of grain types and sizes. The area between grains may be open or filled with finer grains or mineral cement. Foliated or banded metamorphic rocks show a distinctive mineral lineation while massive metamorphic rocks consist of tightly interlocked crystals without lineation. Like igneous rocks, there are no void spaces between grains in metamorphic rock.

a fine-grained basic rock. No quartz is present in any of these rocks.

At about 1,000 to 800 degrees Celsius, the ferromagnesians that form are amphibole (hornblende) and biotite (black mica), and the plagioclase feldspar is now sodium-rich. Minor amounts of quartz also form at this time. Rocks that contain these minerals are intermediate in composition between the basic rocks described above and the *acidic* (or silica-rich) rocks that form at lower temperatures. Consequently they are often called *intermediate* igneous rocks, and common examples are diorite and andesite. They are lighter in color than the basic rocks (usually moderate shades of gray) but not as light as the acidic rocks which are often light gray or white, and sometimes pink or red.

Between 800 to 600 degrees Celsius, the minerals that crystallize from magma are biotite, muscovite (white mica), potassium-feldspar (orthoclase or microcline), abundant quartz, and some hornblende. Granite is a coarse-grained acidic rock that is composed of these minerals. Rhyolite has the same composition, but is fine-grained. The latest stage of cooling may result in pegmatites which are very coarsely crystalline and rich in quartz, potassium feldspar, and muscovite. In some instances during the late cooling stages, pure quartz veins develop.

The rate of cooling influences the size of the crystals that form. If cooling is instantaneous, as would be expected in a thin lava flow, the melt will not have time to crystallize and a glassy rock called obsidian will form. Finely crystalline rocks form in thicker lava flows or in small, near-surface intrusives. Larger masses, and igneous intrusives deep within the crust cool more slowly, and coarsely crystalline rocks result.

If we use the information on the effect of cooling rates on crystal size and the order of mineral formation, we end up with a rock nomenclature system that tells us quite a bit about the formation of that rock. Table I is such a nomenclature system. In this table the rocks in the right-hand column formed at higher temperatures than those to the left. The rocks named across the top crystallized more rapidly than those on each succeeding line.

The sedimentary rocks are deposits of weathered and eroded materials, and many of these are formed of particles derived from the breakup of preexisting rocks. Others represent chemical precipitates of elements dissolved from other rocks. Thus, both particle size and composition are important features to consider in studying sedimentary rocks, just as they were for igneous rocks.

In sedimentary rocks, coarse particles indicate nearness to a source of rapidly eroding rocks, because long-distance

TABLE I
Classification of Some Igneous Rocks

Texture	Composition (Mineralogy)		
	Acidic	**Intermediate**	**Basic**
	Quartz, Potassium Feldspar, Muscovite, Biotite, Hornblende	Little Quartz, Hornblende, Biotite, Sodium-plagioclase	Olivine, Augite, Calcium-plagioclase
Glassy (Minerals not visible)	Obsidian	Obsidian	Obsidian
Finely Crystalline	Rhyolite	Andesite	Basalt
Coarsely Crystalline (Minerals easily visible)	Granite	Diorite	Gabbro

Note: If an igneous rock contains some larger crystals in a finer matrix, the adjective porphyritic is added to the appropriate rock name.

transportation would reduce the fragments to smaller sizes. Coarser grain sizes also require a high-energy transportation mechanism. Steep, rushing rivers or waves crashing on a beach provide such conditions. These transport mechanisms also serve to round the edges of the particles and to sort them by size, shape, and weight.

When rocks weather, the stability of the minerals is determined by the conditions of their original formation. The weathering environment at the earth's surface is one of low temperature and pressure with abundant water, oxygen, and carbon dioxide. These conditions are very different from those which accompany the formation of igneous and metamorphic rocks.

Generally those minerals that form at the highest temperatures and pressures are less stable at the earth's surface, and they weather more rapidly than those that formed at lower temperatures and pressures. This explains why plagioclase feldspar and ferromagnesian minerals break down chemically to produce clay minerals and soluble materials quite easily. As weathering proceeds, resistant minerals like quartz preferentially accumulate along with clay minerals and are available for transport to a depositional site, while soluble materials move in surface and subsurface water toward the sea. When dissolved material reaches the sea, it will precipitate when concentrations become high enough. Evaporation of water in an isolated or restricted part of the sea is a mechanism which increases the concentration of dissolved material to the point where dolomite, (calcium magnesium carbonate), gypsum (calcium sulfate), and halite (sodium chloride or rock salt) will form.

Plants and animals in the sea also take up dissolved material to make siliceous, phosphatic, or calcareous exoskeletons or parts which fall to the sea floor when they die. Corals and many other marine organisms form their protective or supportive structures of calcite (calcium carbonate). The remains of all these can accumulate to form limestone. In fact, almost all carbonate rocks are products of biologic processes.

In ancient seas, magnesium from seawater apparently reacted with recently deposited calcium carbonate and converted it to dolomite, a calcium magnesium carbonate mineral. Rocks composed of dolomite have the same name as the mineral. Dolomite may also form in the tidal zone associated with conditions favoring abundant algal growth.

A classification of sedimentary rocks usually distinguishes between the chemical precipitates (gypsum and halite, sometimes dolomite, possibly some silica as chert) and rocks with a clastic or fragmental texture. This includes those rocks composed of particles of carbonate. In classifying clastic-textured rocks (Fig. 1-4), grain size is used because of its relation to the energy required to move particles about. The use of composition takes into account the basic differences between sea-derived particles (calcite) and land-derived silicic material (quartz, feldspar, and clay minerals). It also allows for an intermediate grouping to account for the mixing of particles with different origins. Such a system is shown in Table II.

Metamorphic rocks are igneous or sedimentary rocks that were texturally or mineralogically changed by heat, pressure, migrating fluids, or movement. Metamorphic alterations that produce massive-appearing rocks include: sandstone and conglomerate to quartzite or conglomeratic quartzite, limestone and dolomite to marble or dolomitic marble, and basalt or greywacke to greenstone. Quartzite is so tightly compacted and cemented with silica that it breaks across the original mineral grains. Marble is often very coarse-grained and is frequently white. Greenstone is a massive, dark rock with a slight greenish cast resulting from the growth of chlorite during metamorphism. All of these examples are denser than the original sedimentary or igneous rock.

Metamorphism involving a directed

TABLE II
Terminology for Some Common Sedimentary Rocks

Grain Size	Composition (Mineralogy)		
	Silicic-rich	Mixed Carbonate and Silicic	Carbonate-rich
Pebbles and Cobbles	Conglomerate	Conglomeratic limestone or limy conglomerate	Limestone conglomerate
Sand-size (2 mm–.06 mm)	Sandstone	Sandy limestone or limy sandstone	Skeletal limestone (fossil debris)
Finer material (less than .06 mm)	Mudstone, siltstone or shale	Argillaceous or silty limestone, calcareous mudstone, siltstone, or shale	Micritic or fine-grained limestone

Notes: If the carbonate mineral is dolomite (calcium magnesium carbonate), substitute dolomite or dolomitic for limestone or limy. Calcareous sandstone implies the presence of secondary calcite cement. Sandstone containing abundant feldspar is called arkose. If it contains minor amounts of feldspar the adjective feldspathic is used. If a sandstone is largely composed of dark rock fragments it is called a greywacke. Shale parts along closely spaced planes essentially parallel to bedding. Siltstone is a grainy rock with particles finer than sand-size.

TABLE III
Terminology for Foliated or Banded Metamorphic Rocks

"Layering"	Grain Size	Chlorite, Mica and Quartz		Amphiboles, Mica, Feldspar and Quartz
Cleavage	Fine	Slate		
Foliated	Medium to coarse		Schist	
Banded	Coarse			Gneiss

Note: Common massive metamorphic rocks are: quartzite, marble, and greenstone.

pressure causes platy or needle-shaped minerals to crystallize in a direction that minimizes the stress (Fig. 1-4). Consequently, these minerals form at right angles to the stress and this results in a crystal parallelism that may look like layering. The rock usually breaks along the direction of preferred mineral alignment. If it breaks with a smooth surface at some angle to bedding, it has slaty *cleavage* and the rock is a slate (Table III).

The aligned grains may be fairly large crystals of micaceous minerals, and then the rock breaks along wrinkled planes on which the sheen of the mica plates is obvious. This type of parting is termed *foliation* and rocks with this property are called schists (Table III). In other instances the minerals are recrystallized dark amphiboles and light-colored feldspars. This gives the rock a light and dark banded appearance and it is called gneiss (Table III). Except for the lineations, such rock resembles granite.

The sequence of rock types from slate to schist to gneiss represents the effects of increased heat and differential pressure applied to the original rock. These progressive changes in rock type represent higher metamorphic ranks or grades. Metamorphic rank can be defined by rock type or by indicator minerals that form during metamorphism. The greenish, micalike mineral chlorite or the dark mica biotite form under low rank conditions. Garnet and staurolite form at the middle rank, while needlelike sillimanite is the indicator for the highest rank of metamorphism.

Rock identification requires that you observe texture, grain size, and mineralogy. The identification of small mineral grains in rocks is not easy, but fortunately there are only 17 common rock-forming minerals. When you learn to identify: quartz, chert, calcite, dolomite, halite, gypsum, potassium feldspar (orthoclase and microcline), plagioclase feldspar, muscovite, biotite, hornblende, augite, olivine, chlorite, garnet, and sillimanite, you are all set.

EROSIONAL PROCESSES

In discussing the rock cycle, the orogenic cycle and the classification of sedimentary rocks, we recognized that weathering and erosional processes moved material from high areas to lower elevations. Thus, high standing areas produced by orogenic events are slowly graded downward. Consequently, erosional processes are often called gradational agents.

The most important gradational agent working to modify the surface of the earth is running water. Running water is part of a system referred to as the *hydrologic cycle* (Fig. 1-5). This system involves all the water of the earth (including glacial ice), most (97%) of which is in the oceans. The sun supplies the energy for this cycle by evaporating water from land and sea. A small additional amount of water is added to the atmosphere by plants (transpiration). Atmospheric water condenses into clouds and under favorable conditions falls as rain or snow. This precipitation is either evaporated back to the atmosphere, taken up by plants, percolated into the ground or runs off to form streams.

Some of the water that percolates into the ground works down to the zone of permanent saturation, where all the spaces between rock and sediment grains are water-filled. The upper boundary of this saturated zone is the *ground water table.* This surface forms a subdued replica of the general topography so that it rises under large hills and is lower under broad valleys. It also fluctuates with the water supply. When there is abnormally high precipitation for a sustained period the ground water table rises. A long drought will cause it to drop. In our region, the water table is usually highest in the spring and late fall.

Where the ground water table intersects the surface of the earth, a lake, river, or swamp will result. Closed, deeper depressions below the water table form lakes while shallow ones form swamps. At the edges of such bodies of water, springs contribute water to the depression. Rivers and major streams also may receive recharge from ground water if they flow at or below the water table. Many of Wisconsin's and Michigan's spring-fed trout streams are examples of this situation.

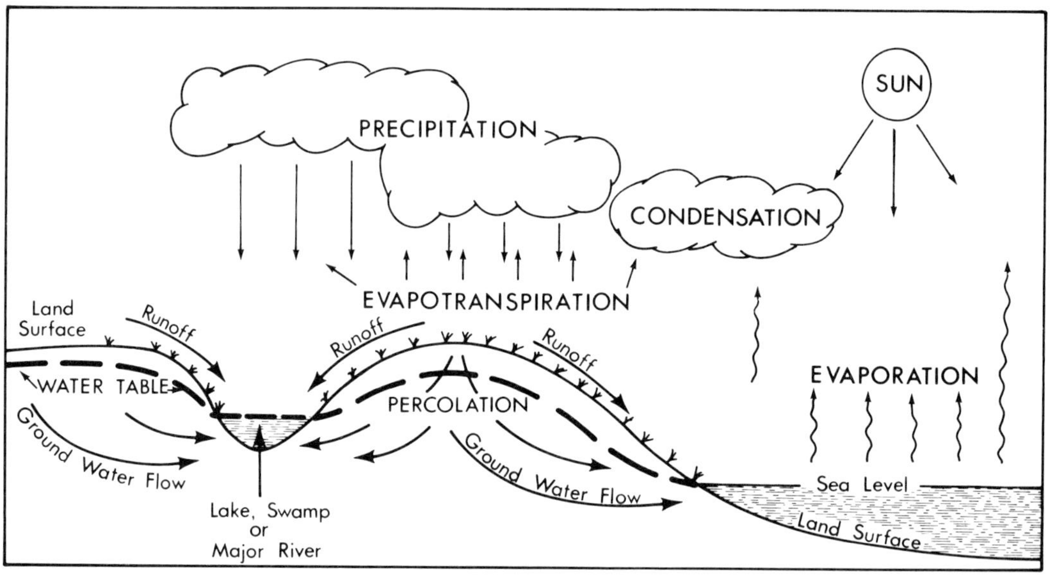

Figure 1-5. Hydrologic cycle illustrating interrelations between atmospheric, surface (seas, lakes, rivers), and subsurface water (ground water).

As water percolates down into the ground, it dissolves soluble mineral matter. If the water is charged with carbon dioxide from the atmosphere it will form a weak acid that dissolves carbonate rock. If sufficient time is available, as it was in that part of our region known as the Driftless Area during the moist Pleistocene, solution caves result.

When water runs off the earth's surface, it seeks the lowest available paths, and soon favored channels develop. These channels enlarge to form streams that remove material as they progressively downcut. A rapidly eroding stream has a steep gradient and cuts a V-shaped channel that ultimately will enlarge into a narrow valley (Fig. 1-6A). As time and erosion proceed, the character of the countryside changes. The streams reduce their gradients and downcutting diminishes. The valleys become wider and slope gently to the stream (Fig. 1-6B). At this stage, the stream begins to wander or *meander* within its valley. Meandering against the valley sides will slowly widen the valley (Fig. 1-6C). Periodic flooding by such streams will inundate these broadened valleys and deposit loads of mud and sand to form a *flood plain* with characteristically rich, fertile soil.

These stages in the history of an individual stream can most easily be appreciated if one follows a given stream from its headwaters in western Wisconsin to where it joins a larger stream, then continues along this larger stream to a river, and ultimately to the Mississippi. The final observations of stream characteristics could be made by following the Mississippi to where its flood plain is many miles across before it builds a delta into the Gulf of Mexico.

As the streams modify their valleys, the upland areas between neighboring streams are also involved. When a region is uplifted, stream downcutting is initially rapid, and valleys are so narrow that the broad interstream areas are not affected. Furthermore, there are very few streams and the drainage of the uplands is poor. As stream development proceeds, the valleys enlarge and interstream areas decrease to become narrow divides. Eventually the entire area is in slope and drainage is well developed. As

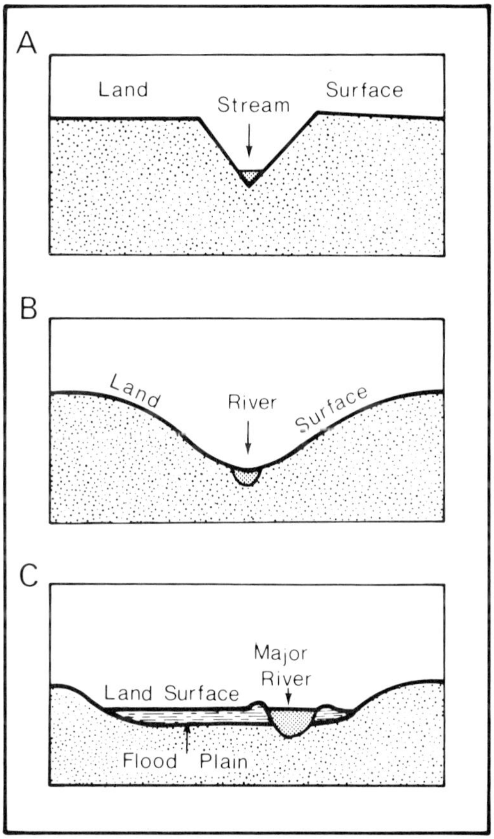

Figure 1-6. Cross sections illustrating a cycle of river erosion. The initial stage (A) of downcutting involves the development of a "V"-shaped valley with relatively flat uplands. As erosion continues, the valley widens and the entire area is in slope (B). Eventually the stream stops downcutting and begins to erode laterally, and a wide valley with a flood plain and low divides results (C).

the cycle continues, the divides will become low and insignificant topographic features.

The precipitation that falls as snow may melt and contribute to the runoff, percolation, and evaporation. However, under cooler conditions than we presently have in this area (maybe as little as 5 degrees Celsius in our average annual temperature), the snow that falls wouldn't all melt from year to year. If this happened, annual accumulations of snow would remain and their weight would cause recrystallization into ice. If the ice develops to a sufficient thickness, a glacier forms and gravity causes it to flow outward from its highest elevation (Fig. 1-7). Where melting rates exceed accumulation, the glacier will cease to grow and it will be reduced in size. At present there is more water in the world that is tied up in glacial ice (2%) than there is in the atmosphere, lakes, streams, swamps, and the ground.

Advancing glacial ice is a powerful gradational agent that is capable of eroding and transporting enormous quantities of rock, some of it as extremely large boulders. This transported material is deposited in a variety of situations that will be described later. Evidence as to the effectiveness of glacial erosion and deposition in modifying the landscape is common in this area.

The sun also provides the energy to drive atmospheric circulation patterns. These systems result in the winds that do geologic work either directly, or as wind-generated waves and currents in lakes and oceans.

Wind alone is not a strong erosional agent and can only move material up to sand-size. However, where the wind is strong and constant and a sand supply is available, dune fields such as those formed along Lake Michigan at Wisconsin's Terry Andrae State Park, Illinois Beach State Park, and the Indiana Dunes National Seashore will form. Similar conditions existed around the shore

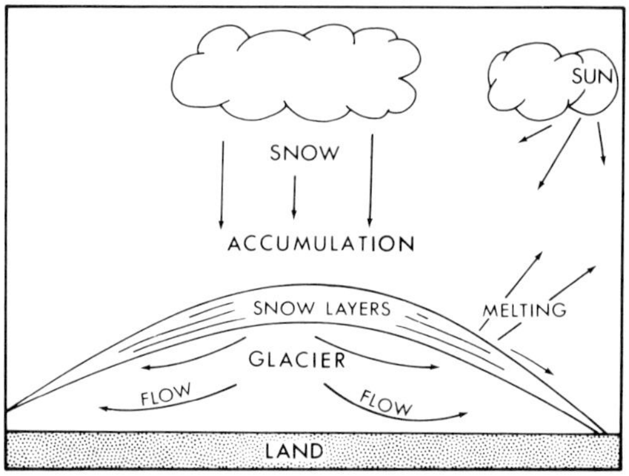

Figure 1-7. Accumulating snow compacts and recrystallizes into glacial ice that flows outward because of its own weight. When melting exceeds accumulation, the glacier will diminish in size and thus appear to retreat.

of Glacial Lake Wisconsin and along higher lake levels of ancestral Lake Michigan where "fossil" dune fields still exist to mark the edges of these old water bodies.

Waves generated on water move as wave forms away from the wind until they meet the land. The water involved in the waves moves in an orbital pattern in one place as the individual wave forms move forward through the water. When a wave reaches shallow water, the orbital path of the water "feels" the bottom and drags against it. This causes the water at the surface to move forward and a breaker results (Fig. 1-8). When a wave feels the bottom, breaks, and plunges over, it scours an erosional trough. The material removed from this trough is carried onto the beach by the run-up from the breaking wave. It may also be moved seaward by retreat of the run-up from the beach. This action builds up the beach onshore, while offshore a bar may develop. It is easy to see this activity going on along a beach if you throw small, colored fragments of tile or brick into the surf.

Often waves are directed toward the land by a wind that is at some angle to the shoreline. When this happens the waves in shallow water slow down when they "feel bottom," while they continue their open-water rate of advance in deeper water. The result is that the wave fronts swing more nearly parallel to shore as they are progressively retarded in the shallower water. As each advancing wave breaks, it forces retreating water from the previous wave run-up to move along the shore (Fig. 1-9). This generates a *longshore current.* As the water moves, so do the sand particles (Fig. 1-9). Each grain tends to trace a series of semielliptical paths as it travels parallel to the beach under the influence of the longshore current.

If a bay with deeper water is encountered, the longshore current disperses. The sand in transport will then be deposited to form a *spit* parallel to the general trend of the shoreline (Fig. 1-9). If a spit bends sharply shoreward due to wave action, it is called a *hook.* Sometimes spits will grow completely across bays producing *baymouth bars.*

The hydraulic impact and abrasional power of waves against the land produce a variety of erosional features. Since the energy of waves is usually focused on projecting headlands along a shoreline, these features are preferentially eroded. Steep cliffs, pillars of rock, caves, and arches develop along bedrock headlands where wave erosion is active.

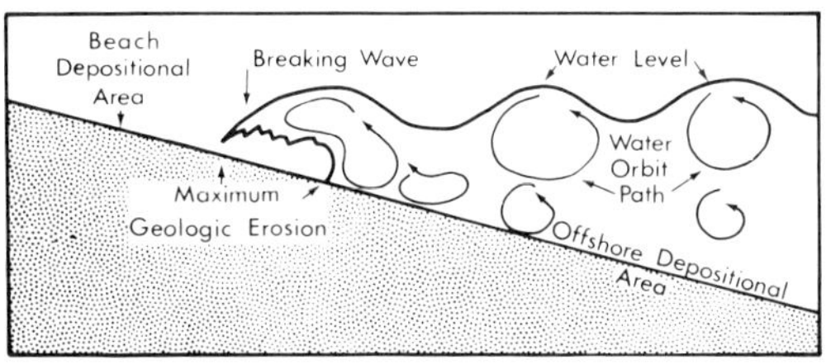

Figure 1-8. Orbiting paths of wave forms in water are distorted by the bottom as they come ashore. This causes the top part of the orbiting water mass to fall forward as a breaker. Scouring occurs at the breaker line, and there is deposition of eroded material seaward and landward from this developing trough.

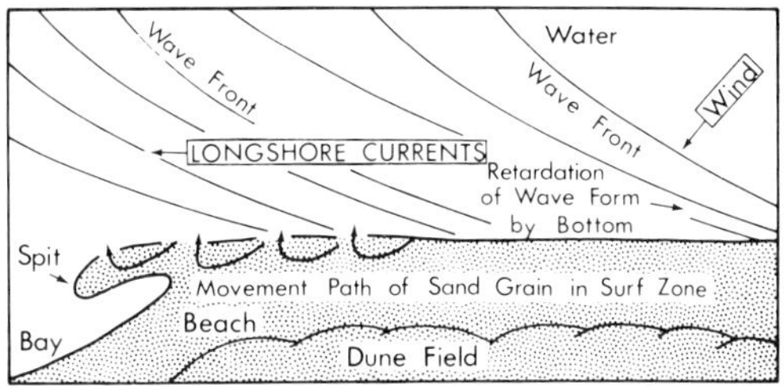

Figure 1-9. Development of longshore current by wave fronts approaching the beach at an angle. The movement of sedimentary particles along the beach in response to the longshore current is shown. If a current moving along the shore enters deeper water it loses energy, and sediment in transport will be deposited to form a spit.

GEOLOGIC TIME

When we discussed the rock and orogenic cycles, it was apparent that they involved sequences of events over long periods of time. The concept of uniformitarianism was later applied to erosional processes operating in the geologic past, and personal experience told us that these events are generally slow in terms of major changes on the face of the earth. Even with such feelings, it is difficult, if not impossible, for each of us with our short individual life spans and the relatively brief historical record of our species, to fully appreciate the significance of geologic time.

The earth is believed to be five billion years old and the oldest rocks dated on earth are some 3.8 billion years old. If we lived for 100 years, it is obvious that a lot of geologic events would occur during this time. There would be volcanic eruptions, earthquakes, landslides, floods, erosion of high standing areas, changes along the shorelines of lakes and seas, etc. Even then, we wouldn't see glaciers cover part of North America or seas inundate most of the world's land surface. But if we had lived for 3,800,000,000 years we would have seen these things and many more.

We know the glaciers began to retreat from the uppermost Midwest more than 10,000 years ago and the many distinctive features they formed remain for our inspection today. We can also find marine fossils in the sedimentary rock layers of Wisconsin and adjacent states that testify to an event that happened before the glacial deposits formed because they are underneath them. This reasoning involves the concept of *superposition,* which dictates that the depositional record of the youngest event must overlie previous depositional records. A very local and specialized exception to this rule would be deposits forming in caves within existing rocks. It is also obvious that the law of superposition does not work for intrusive igneous rocks, although it applies to extrusives such as ash falls and lava flows.

Superposition allows us to tell relative geologic time in layered rocks. To read geologic history in such rocks, we look at the layers from the bottom up since this is the progression from the earliest to the latest event. An example of a simple progression of events is shown in Figure 1-10A. If the sequence in Figure 1-10A should be folded by compressional forces, then uplifted and eroded, it would resemble Figure 1-10B. The history is carried further in Figure 1-10C, when a sandstone bed containing fragments of the underlying bedrock is deposited. Before the sandstone formed, this area was eroded, but no record of geologic events during this time interval remains since there was no deposition at this locality. A nonrecord such as this is called an *unconformity.* These are gaps in the geologic record similar to a missing chapter in a history book. The erosional gap in Figure 1-10C is called an *angular unconformity* since the underlying rocks are folded or tilted and meet the overlying deposits at an angle.

The base of the uppermost sandstone bed in Figure 1-10C contains material reworked from below the unconformity. This type of deposit is called a *basal conglomerate,* and the fragments of underlying rock within the sandstone proves that this deposit formed after the underlying rock. This verifies the law of superposition and serves to illustrate the *law of inclusions.* This law states that the enclosing rock is younger than the included material, and it also applies to igneous rocks.

The sequence in Figure 1-10C is cut by a dike or plug that feeds a lava flow at the surface. We know this intrusion is younger than the unconformity and the overlying sandstone because an igneous body can't cut through something that isn't there. This is the *law of cross-cutting relationships.* The extruded lava overlies the sandstone so the law of superposition also shows it to be younger. Sandstone inclusions in the dike or the lava would also establish the relative age relations.

We could carry the history of Figure 1-10 on by depositing more sediments on the cooled and eroded lava, or we might fold, uplift, and erode the entire sequence once again. Events such as this have happened throughout geologic time in many parts of the world, and the record is there in the rocks for us to read.

Once we can read geologic history in rock sequences, as illustrated in Figure 1-10, we can understand the development of a relative geologic time scale. In Figure 1-10 we put geologic events in the proper order, and this

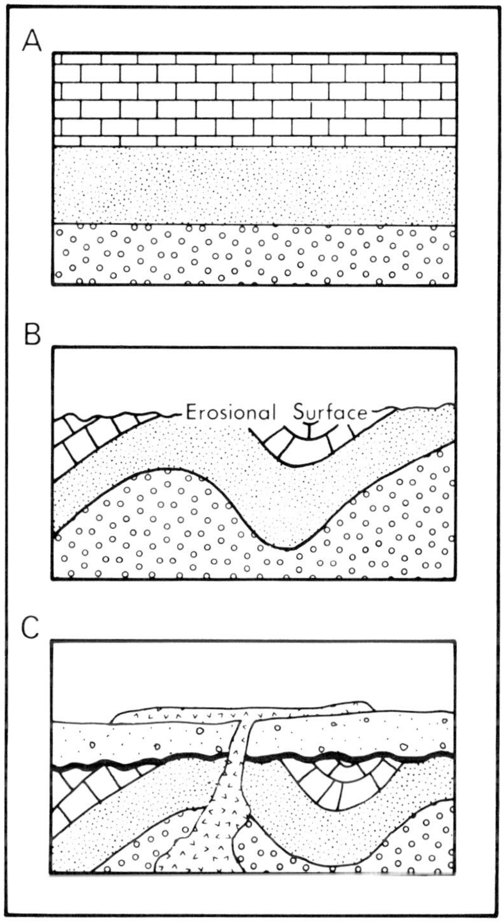

Figure 1-10. The upper diagram shows a sequence of geologic events as they might be recognized on rock exposed on a quarry wall or a road cut. The layer representing the earliest event is at the bottom and the symbols represent deposition of pebbles; possibly on a beach or in a stream. The layer representing the next event is indicated by the stipple pattern that is the usual symbol for sand or sandstone. This type of deposit might form on a beach, in a stream, or as a sand dune. Overlying this is the blocky pattern representing limestone, that probably formed in a shallow sea. The presence of marine fossils in the limestone would prove this speculation.
The middle diagram shows the sequence described above as it might look if it were folded by compressional forces, uplifted and eroded. The limestone unit is almost gone.
The bottom diagram shows a newly deposited sandstone bed that contains fragments reworked from the underlying bedrock. This sandstone, and the underlying sequence, are then cut by an igneous dike that reaches the surface to form a lava flow.

established the time sequence for these events. Relative rock ages are simply determined if we remember the laws of superposition, inclusions, and cross cutting relations.

Another way to unravel the sequence of geologic events and to appreciate the incredibly long span of geologic time is to use the concept of faunal and floral succession. This biologic approach to geologic time can be made in two ways. One way is to observe the plant and animal fossils you find in each layer of a sequence, such as shown in Figure 1-10A. We don't need to know what they are called or what family of plants or animals they represent. Once we have a biologic succession worked out in one area, we can go elsewhere and try to find the same forms. If we are successful, we are comparing (or correlating) fossils and consequently the rocks which contain them by a purely empirical approach.

Correlation works reasonably well, but especially so in local areas. However, we know that different faunas and floras do not always reflect rocks of differing geologic age. A change in fauna and flora may also reflect differences in the depositional environment. For example, there is no reason to expect the fossils in a beach deposit will be identical to those that accumulated offshore in deeper water, since the living forms are not the same in these different environments in our modern seas. Once again, uniformitarianism aids geologic understanding.

When Charles Darwin's *Origin of the Species* was published in 1859, the empirical approach to faunal and floral succession became blended with evolutionary concepts. By using collections from rock sequences, even a novice can begin to see a pattern of change in similar types of fossils. Both the type and diversity of fauna and flora for the various layers will show change with time. Generally speaking, the complexity of individual forms and the diversity of the total fauna and flora increase from older to younger rocks. The fossilized record of certain forms may abruptly end while others suddenly seem to start at various places in the rock succession.

As fossil collections accumulated and were studied in detail, some well-defined evolutionary patterns were established. These allow us to use individual fossils to place rocks in a relative time sequence. Many fossil forms attained worldwide distribution so this enables us to relate rock sequences (and relative time lines) across the oceans. Once this was done, it became possible to define a sequence of geologic history for the entire earth. This was a great step beyond our ability to merely interpret the geologic events in relatively isolated areas, such as illustrated in Figure 1-10.

Geologists, however, were not satisfied with a concept of relative geologic time. They computed the rate of sediment accumulation in the modern ocean and then measured rock thickness in order to estimate actual years required for various types of sedimentary sequences. They also measured the salt derived from the weathering of rocks over a short span of time in order to determine how many years it would take for freshwater seas to reach the salinity of present-day seawater. These attempts all yielded very young ages when compared to concepts of evolutionary rates of change in fossil plants and animals as they were known from life on earth.

Ultimately atomic science and instrumentation progressed to a point where certain types of minerals and some bone and woody materials could be age-dated by known rates of atomic "decay" of elements such as uranium, potassium, rubidium, and carbon. This knowledge has proved especially helpful in dating a sequence of orogenic cycles in areas with complex histories.

The technique of determining absolute age consists of applying knowledge of the rate of change from heavy, relatively complex and unstable atoms or isotopes to atoms or isotopes that are lighter, simpler, and more stable. The rate of these changes is unaffected by all but the highest temperatures and pressures found within the earth's crust. However, high rank metamorphism causes these atomic "clocks" to be reset to the time of the metamorphic event.

Accumulation of decay products from a radioactive element starts when it is included within a mineral crystallized from a magma. If we date the time of formation of the mineral from the amounts of the element and its decay products, we effectively date the mineral, the rock containing the mineral, and the igneous event in years before the moment of the actual analysis. Consequently absolute ages are termed "before present" (BP). Since we know how many igneous events fit into the relative geologic time scale, we can also date related sedimentary events quite closely. As absolute ages become available from all over the world, we can at last put years BP on the relative framework that evolved over some 200 years of geologic study. The geologic time scale

TABLE IV
Geologic Time Scale

Modified from Harbaugh (1968). Estimates of time are modified from Holmes (1965), Kulp (1961), and others. There is a margin of error in these dates that will improve as age-dating techniques are refined and are applied more widely.

Era	Period	Epoch	Millions of Years Ago	Duration in Millions of Years	
Cenozoic	Quaternary	Recent			
		Pleistocene	0-2	2	2
	Tertiary	Pliocene	2-13	11	
		Miocene	13-25	12	
		Oligocene	25-36	11	
		Eocene	36-58	22	61
		Paleocene	58-63	5	
Mesozoic	Cretaceous		63-135	72	
	Jurassic		135-181	46	167
	Triassic		181-230	49	
Paleozoic	Permian		230-280	50	
	Pennsylvanian*		280-310	30	
	Mississippian*		310-345	35	
	Devonian		345-405	60	370
	Silurian		405-425	20	
	Ordovician		425-500	75	
	Cambrian		500-600	100	
Precambrian:	No standardized divisions for the Precambrian exist, although various local classifications are used. By definition, the Precambrian embraces the time between the origin of the earth, and the beginning of the Cambrian Period. The oldest rocks dated on earth are about 3800 million years old. However, certain meteorites have yielded age dates in the range 4500 to 4800 million years and this suggests a maximum age for the earth and other planets of about 5000 million years. Thus the Precambrian represents about 88% of geologic time.				

*Together the Pennsylvanian and Mississippian are referred to as the Carboniferous.

with such dates is provided in Table IV. It should be noted that the eras, periods, and epochs were established before absolute age-dating and are not of equal time duration. It is also apparent that older eras and periods represent longer spans of time.

TYPE SECTIONS AND CORRELATION

In our discussion of relative geologic time we talked about a sequence of rocks where the individual units varied in rock type through a vertical sequence from conglomerate to sandstone and then limestone (Fig. 1-10A). When we discussed sedimentary rocks, we noted that each rock represented a response to the materials available for deposition and the environment at the depositional site. Conglomeratic rocks imply a nearby, probably high-standing source and a depositional site with high energy to spread the material around. Rivers or beaches are examples of suitable environments for conglomerate deposition.

Finer materials such as shales imply a distant, possibly low-lying source and a quiet depositional site. Carbonate rocks (limestone or dolomite) imply the lack of a land source area. Furthermore, carbonate rocks reflect the presence of abundant carbonate-secreting marine invertebrates and/or algae. Favorable conditions for growth of these organisms include warm, relatively shallow marine waters that are free from mud and sand. Figure 1-11 is a diagram illustrating sediment deposition as it relates to energy levels, distance from source and water depth.

Figure 1-11 also shows how sedimentary rocks vary laterally. In Figure 1-12 the effect of a continuing marine transgression is illustrated, and the landward shift of the coarser rock types results in a fine over coarse vertical succession of sedimentary units.

The variation in sedimentary rocks both laterally and vertically is easy to visualize in diagrams and on large-scale rock exposures such as one might find on steep mountainsides. However, when smaller, often isolated, exposures are all that are available, we must relate these bits and pieces of information in order to interpret geologic events such as we did for Figure 1-10C. This is done by comparing fossils between two rock ex-

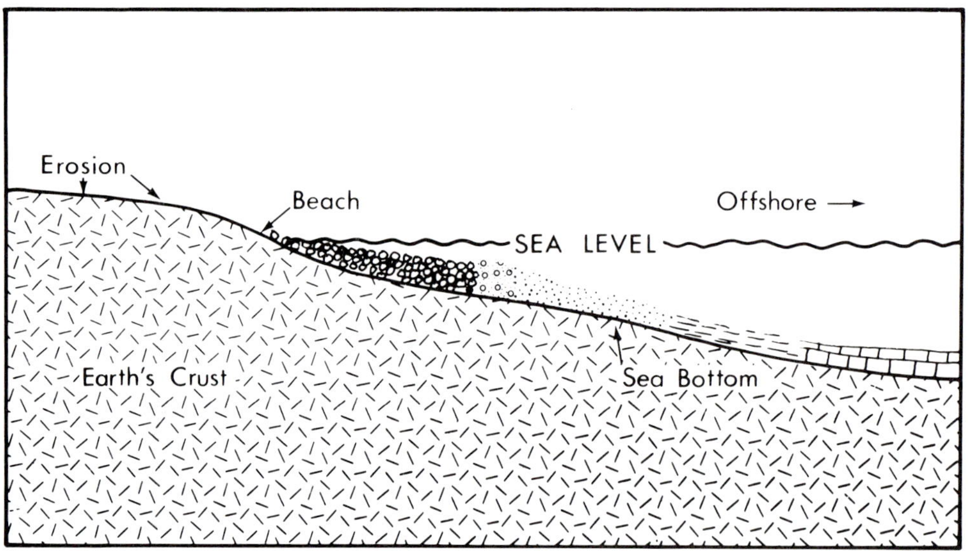

Figure 1-11. A cross section illustrating an eroding land mass which is the source for materials washed into the sea. The coarsest land-derived materials are deposited in the high energy beach environment. This gravel deposit grades out to sand-size particles and then to fine muds. Where the land source is ineffective, limestone accumulates.

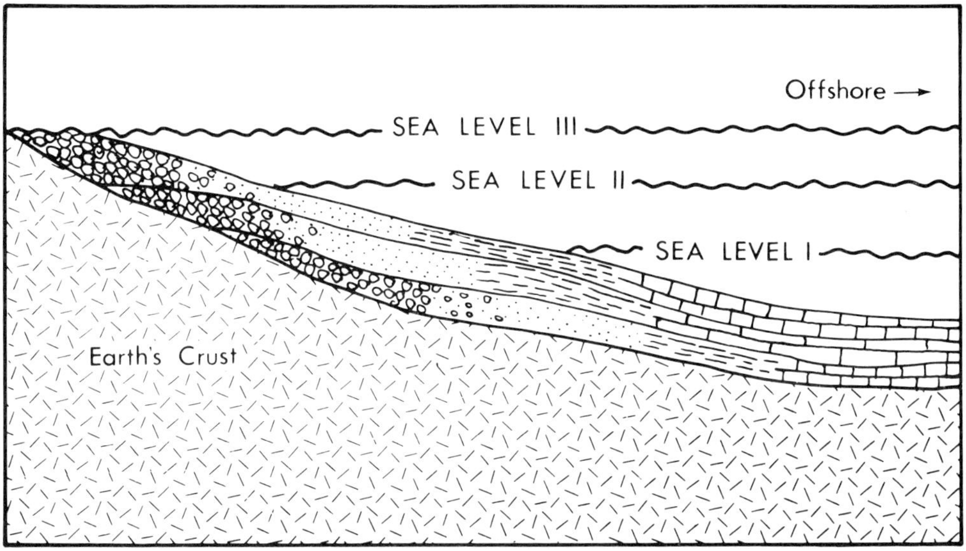

Figure 1-12. A cross section illustrating transgression of the sea so that it stands at three progressively higher elevations on the land mass. As the transgression proceeds landward, fine sediments overlie coarse sediments. The reverse situation would occur if the seas slowly retreated (regressed) from the land. The sedimentary sequence would then be coarse over fine.

posures or by correlating the rocks by comparing similar lithologies or similar sequences (Fig. 1-13).

The characteristics of a rock unit which has been carefully studied, measured, and described in a given place can usually be recognized over a fairly wide region. Normally each recognizable rock unit is named for a geographic feature near the area of its characteristic development. This geographic name then becomes the formational name and a nearby measured vertical sequence of the rock is designated as the *type section.* The use of geographic names for rocks has proved to be a convenient and precise way for geologists to identify a particular unit.

The name applied at the type section is extended outward from this area by careful correlation as far as the distinctive characteristics of the rock unit can be recognized. When the distinctive characteristics of a formation are no longer recognizable, another type section and a new name is established. In Fig. 1-11 we see that 4 names are required to handle the lateral changes in lithology; one each for the conglomerate, sandstone, shale, and limestone. Each then has a name that imparts a clear picture of lithologic characteristics to one who knows the type section from studying literature or from personal observation.

Igneous and metamorphic rocks are also named by this procedure. However, the techniques of correlation described in Fig. 1-13 do not apply to intrusive igneous rocks and high grade metamorphic rocks.

STRUCTURE

When sedimentary rocks are deposited as shown in Figure 1-12, it is apparent that they are not laid down as perfectly horizontal

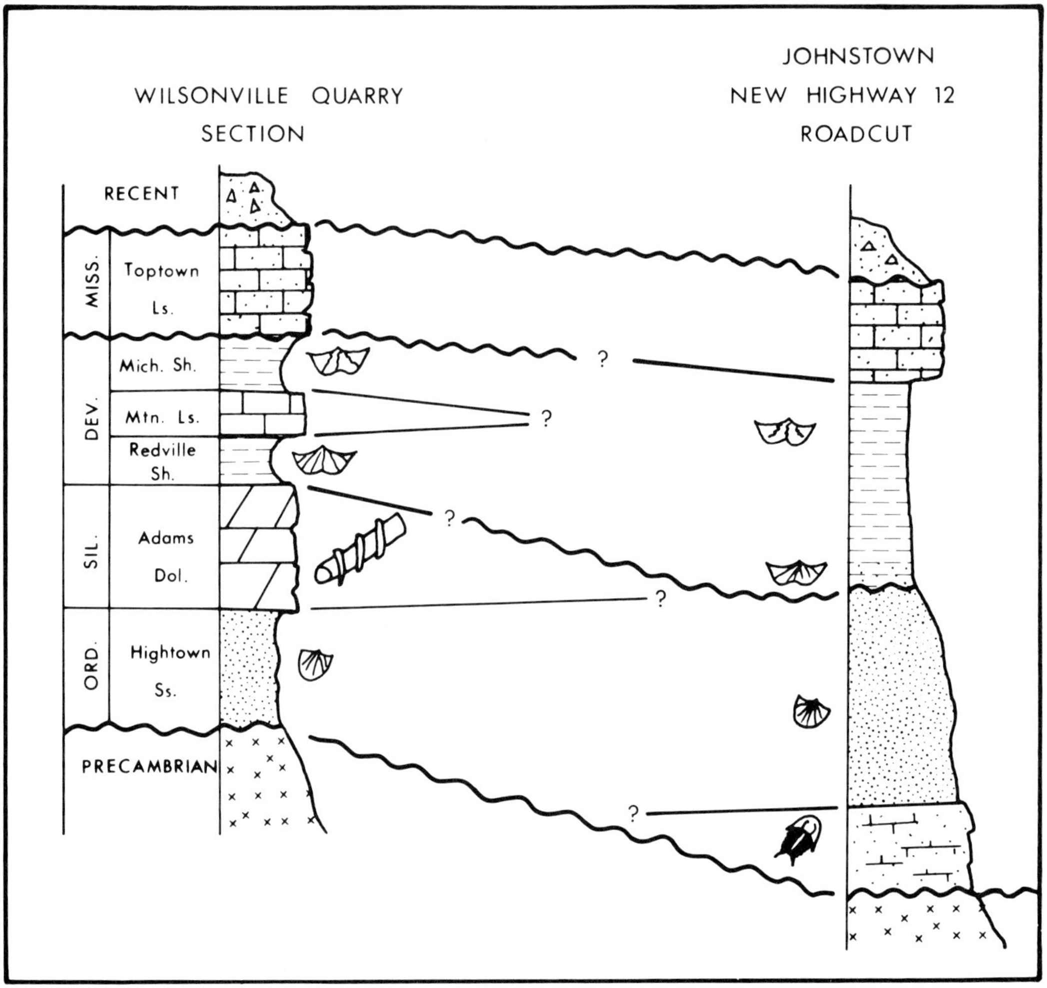

Figure 1-13. A hypothetical correlation diagram between a well-studied, dated, and named rock sequence in a quarry, and a new exposure along a roadcut. Three unconformities are known in the quarry section (the wavy lines). The unconformities between Precambrian and Paleozoic rocks, and Paleozoic rocks and Recent soils were easy to recognize at the roadcut. Similar fossils established correlation of the Ordovician and the Devonian. The lithologies were also similar but the Devonian limestone unit present in the quarry was not deposited in the roadcut section so the correlation lines show it thinning between these two localities. A new basal unit with distinctive fossils is present in the roadcut and the correlation line shows it ending against the Precambrian unconformity somewhere between the two sections. The Silurian Adams Dolomite, with its distinctive fossils, is not present at the roadcut and its removal along an unconformity somewhere between the two sections is indicated by the correlation lines. Correlation of the Mississippian unit is based on its position above the Devonian shales, and a similarity in lithology.

layers. We also know from oceanographic studies that most marine sediments are deposited with slopes of less than 3 degrees. However, there are sedimentary rocks such as certain beach sands and the flank deposits of coral reefs with abnormally high depositional dips. In spite of this knowledge, we still use the *law of original horizontality* which requires us to consider most sedimentary rock units as originally flat-lying deposits.

Once we accept the law of original horizontality, it follows that any deviation in the inclination of the bedding planes of sedimentary rocks from the horizontal results from forces operating within the earth's crust. There are many forces that deform the crust but they generally act on rocks in a vertical or horizontal direction. Vertical forces may be directed up or down while horizontal forces are compressional (a pushing together) or tensional (a pulling apart).

Rocks react to compression and tension in two basic ways; they either fold or break. How a rock yields depends on the rate of application of the force and the physical character of the rock itself. Rapid applications of force usually result in breakage. Hard, well-cemented rocks like quartzite tend to be brittle and they often break when forces are applied. Rocks like limestone that recrystallize easily will often fold under similar circumstances. However, deep burial with its increasing heat and pressure changes the characteristics of rocks, and even those that are brittle at the surface will respond to forces at depth by either folding or some form of flowage.

Common responses of rock sequences to compressional and tensional forces are summarized in Figure 1-14. These structures are illustrated on block diagrams so that two vertical cross sections and a map view are visible. It is obvious that these structures don't appear the same from different perspectives. This illustrates that geology is a three-dimensional science. From usually limited exposures, an observer must try to visualize the three-dimensional aspect of structures.

In describing structure in rocks, a geologist uses measurements of strike and dip. *Strike* is the line defined by the intersection of an imaginary horizontal plane with an inclined rock surface such as the top of a sedimentary rock unit or a fault plane. It is determined by holding the edge of a leveled, straight-sided compass against the surface to be measured. The bearing of the strike is then read off the compass dial relative to true north. *Dip* is the maximum angle of inclination of the surface from horizontal. This measurement is made at right angles to the strike line by an inclinometer graduated in degrees. It is also necessary to record the general direction of the dipping surface.

Strike and dip define a dipping planar surface by providing us with the sense of direction and the amount of inclination. These measurements are also easy to illustrate on maps. The strike is plotted on the map at the place where the observation was made as a short line segment oriented in the proper compass direction. The dip is illustrated by a shorter line projecting at a right angle from the center of the strike line in the proper direction. The amount of dip is recorded next to the dip and strike symbol in such a fashion: ⊤25. In this way, we can plot and read the third dimension on a two-dimensional map.

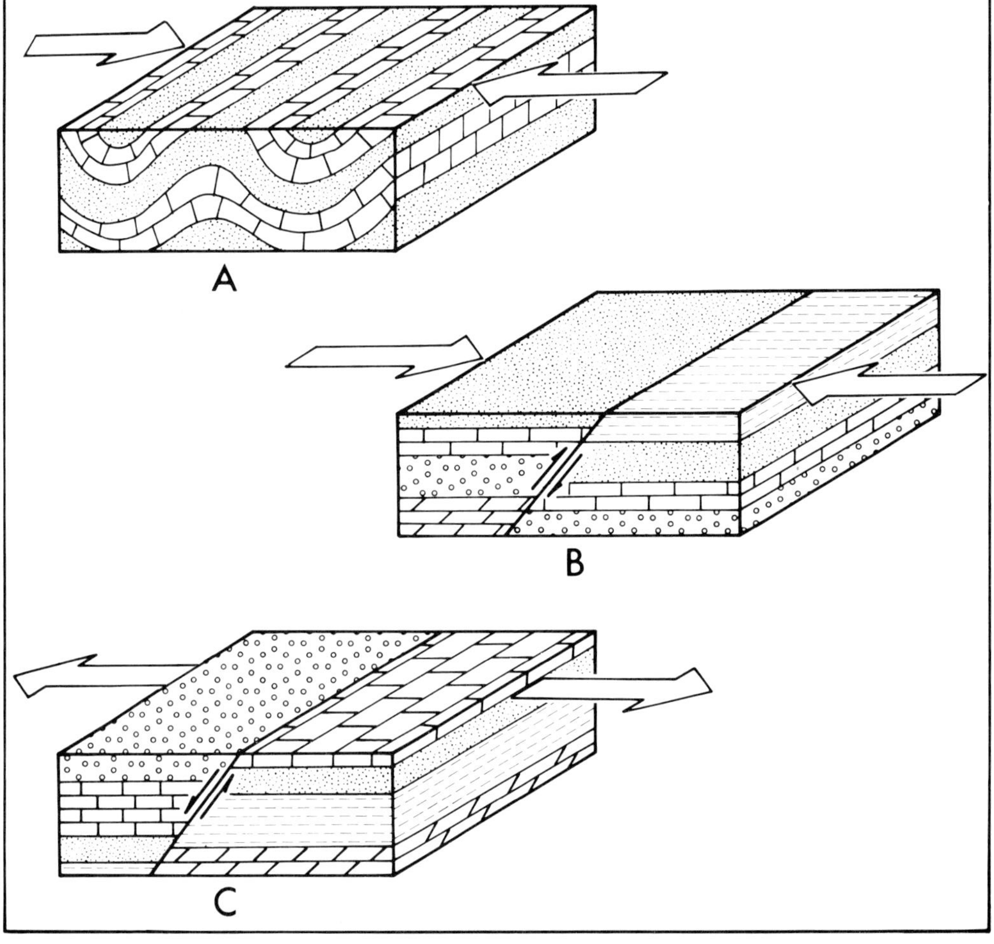

Figure 1-14. Blocks A and B show different responses to compressional forces suggested by the large arrows. Block A illustrates rocks that have folded into *anticlines* (the arches or upfolds) and *synclines* (the troughs or downfolds). The plan or map view shows that the older rocks are in the center of the anticline while younger rocks are exposed in the center of the synclines. Block B shows a compressional fault that is usually called a *reverse fault.* If the fault plane of a reverse fault is inclined at 25 degrees or less it is called a *thrust fault.* In reverse and thrust faults the older beds move up along the inclined fault plane relative to the rocks underlying the fault plane (note the small directional arrows along the fault plane). In contrast is the tensional fault shown in Block C, with tensional forces illustrated by large arrows. This type of fault is called a *normal fault.* The movement on such faults (again, note the small directional arrows along the fault plane) has younger beds moving down along the inclined fault plane relative to older beds below the fault plane.

Chapter 2

GEOLOGIC HISTORY

REGIONAL SETTING

The surface of the earth as we observe it today represents a complex of events that began with the origin of the earth some 5 billion years ago. Our purpose is to examine these events as they relate to northern Michigan, Wisconsin, and adjacent states, based on actual observations you can make in this area.

Our ability to look back in time and successfully unravel the geologic history of any given area is dependent upon the data available. The basic information is the visible rock record. An area with abundant rock exposures and numerous drilled wells or deep mines will obviously tell us more than one that is covered with soil or where no subsurface records are available. Similarly, the more geologic study in a given area, the more we know. We are fortunate because most of the rock exposures in this area have been studied by geologists in sufficient detail to provide us with a basic understanding of geologic history.

The generalized geologic setting of the north-central states, shown in Figure 2-1, is within the interior of the North American continent. The area includes parts of two major physical provinces: the southernmost part of the Canadian shield and the stable continental interior. Wisconsin and upper Michigan lie astride these two vast regions of different geologic character. Both provinces have remained free of orogeny for more than a billion years, in contrast to the more recently active continental margins.

The Canadian shield, that projects into the northern part of this area as the *Superior Upland Province,* is mainly composed of metamorphic and igneous rocks of Precambrian age, greater than 600 million years old. These old rocks were intensely folded, faulted, and subjected to igneous activity during a long and complex history.

In contrast, the bedrock of the stable interior of North America consists of nearly flat-lying or gently folded sedimentary strata of Paleozoic or younger age, that generally thicken outward from the shield toward the unstable continental margins. As shown on Figure 2-2, the northern part of the stable interior is called the *Central Lowlands Province.* This region of moderate topography stretches across the midcontinent from the Appalachian Mountains to the Great Plains. To the south, this province is bounded by the *Ozark Plateaus* of Missouri and the *Interior Low Plateaus* of Kentucky and Tennessee. Both the Central Lowlands and the Superior Uplands were affected by continental glaciation during the Pleistocene but the southerly plateau provinces were not.

The geology of Wisconsin and upper Michigan includes two basically different areas: an area of Precambrian rocks to the north, and a region with Paleozoic sedimentary rocks that overlie the older rocks on the south, east, and west (Fig. 2-3). Since we interpret geologic history from the oldest event

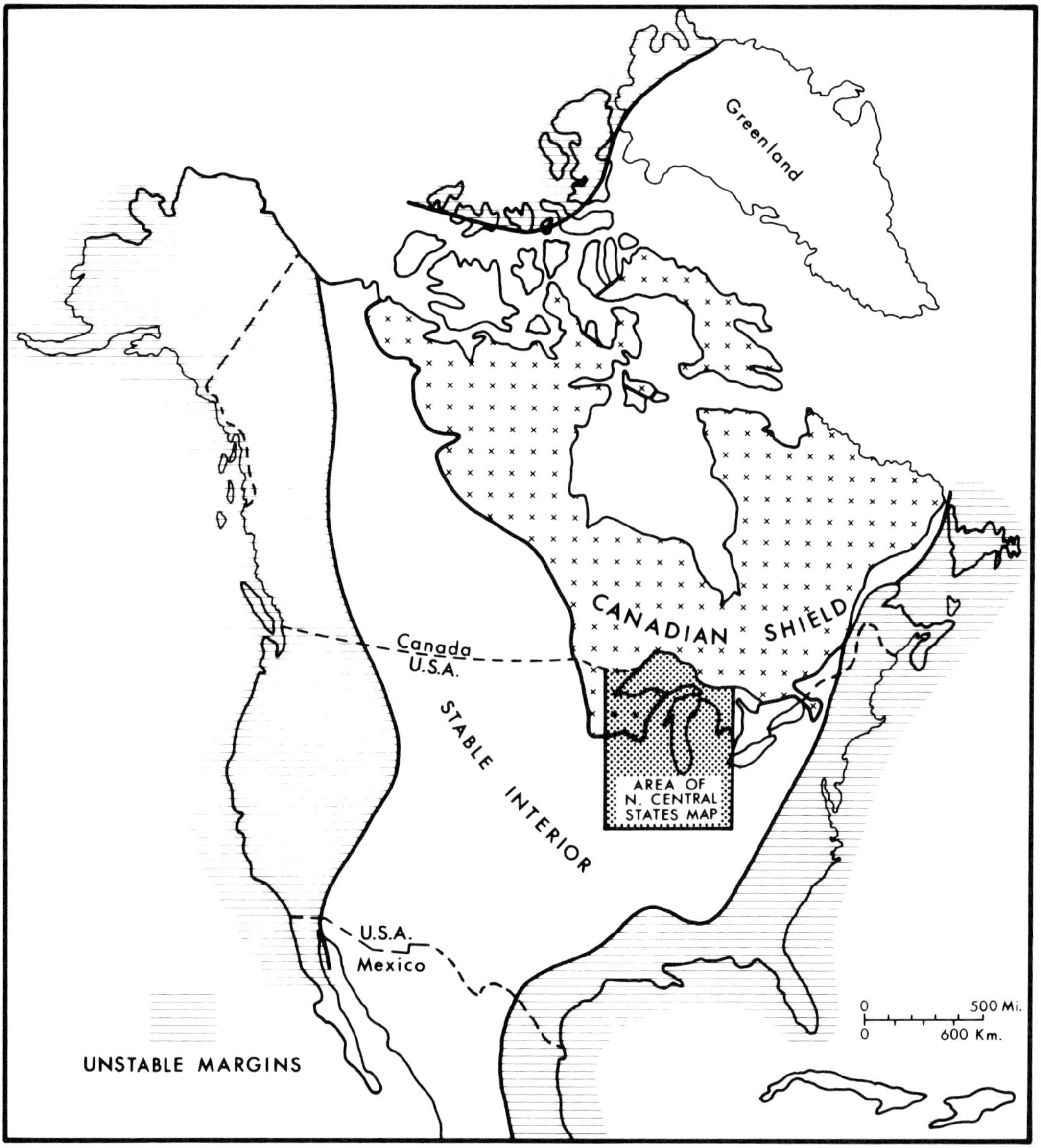

Figure 2-1. Generalized geologic setting of the north-central states. The area shown includes part of the Precambrian Canadian shield and the stable continental interior. The North American continent is flanked by unstable margins.

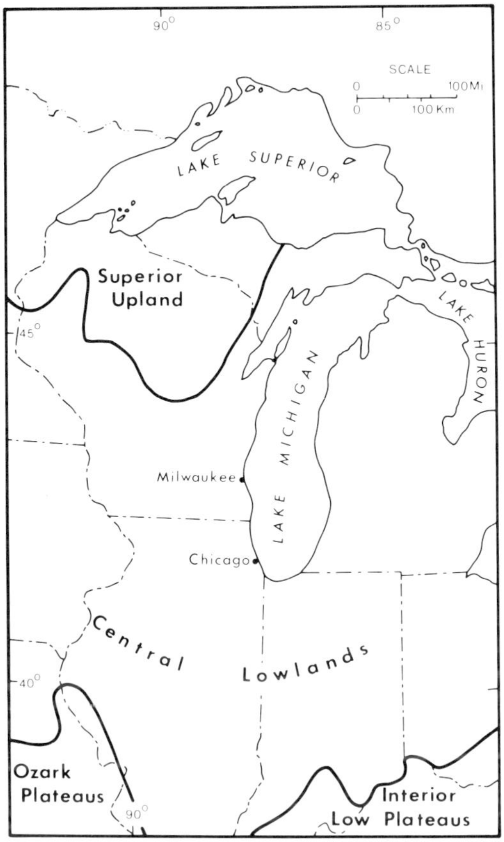

(AFTER FENNEMAN 1938)

Figure 2-2. Physical provinces of the north-central United States. The Central Lowlands are bounded on the north by the Superior Upland and on the south by plateau provinces.

to the youngest, we will start our discussion with the most ancient Precambrian rocks known in our area.

PRECAMBRIAN: MOUNTAIN BUILDING, VOLCANOES AND EARTHQUAKES

Deciphering Precambrian history is a difficult task since earth processes have operated on these rocks for billions of years. Each tectonic event that deforms or metamorphoses progressively younger rocks also involves the older, underlying rocks. Therefore, to understand earliest Precambrian history we try to remove the overprint of all later events. Furthermore, our record of these early events is limited because the law of superposition dictates that older rocks underlie all younger rocks.

In spite of the shortage of exposures and the structural and metamorphic complexity of these early rocks, geologists have been quite successful in developing a sequence of events for Precambrian rocks in the southern part of the Canadian shield. This history involves sedimentation in ancient oceans, volcanic activity, mountain building, intrusion of large masses of igneous rocks, and periods of extensive erosion. The development of this history reflects the continuing efforts of geologists involved in exploiting the economic deposits of copper and iron found within Precambrian rocks. However, since most of this economic interest in the past has been confined to the Upper Peninsula of Michigan and northernmost Wisconsin, there is still much to learn about the Precambrian of Wisconsin that is largely covered by Pleistocene glacial deposits.

Early Precambrian

The oldest rocks exposed in our area are the Keewatin greenstones of Marquette County, Michigan.[1] Similar rocks may occur in Iron County, Wisconsin and possibly in a few other areas (Fig. 2-4). These rocks are

1. Most of us are lucky if we know the names of the counties that surround the one we live in. Consequently we have provided a map with the names of all the counties in the region we are describing in Figure 2-6. The highways and trips shown on this map are those that are described as geological field excursions in the companion volume to this book. Unfortunately this map doesn't contain the names of the communities and parks we mention. For these, we urge you to acquire an indexed road map and use it whenever necessary.

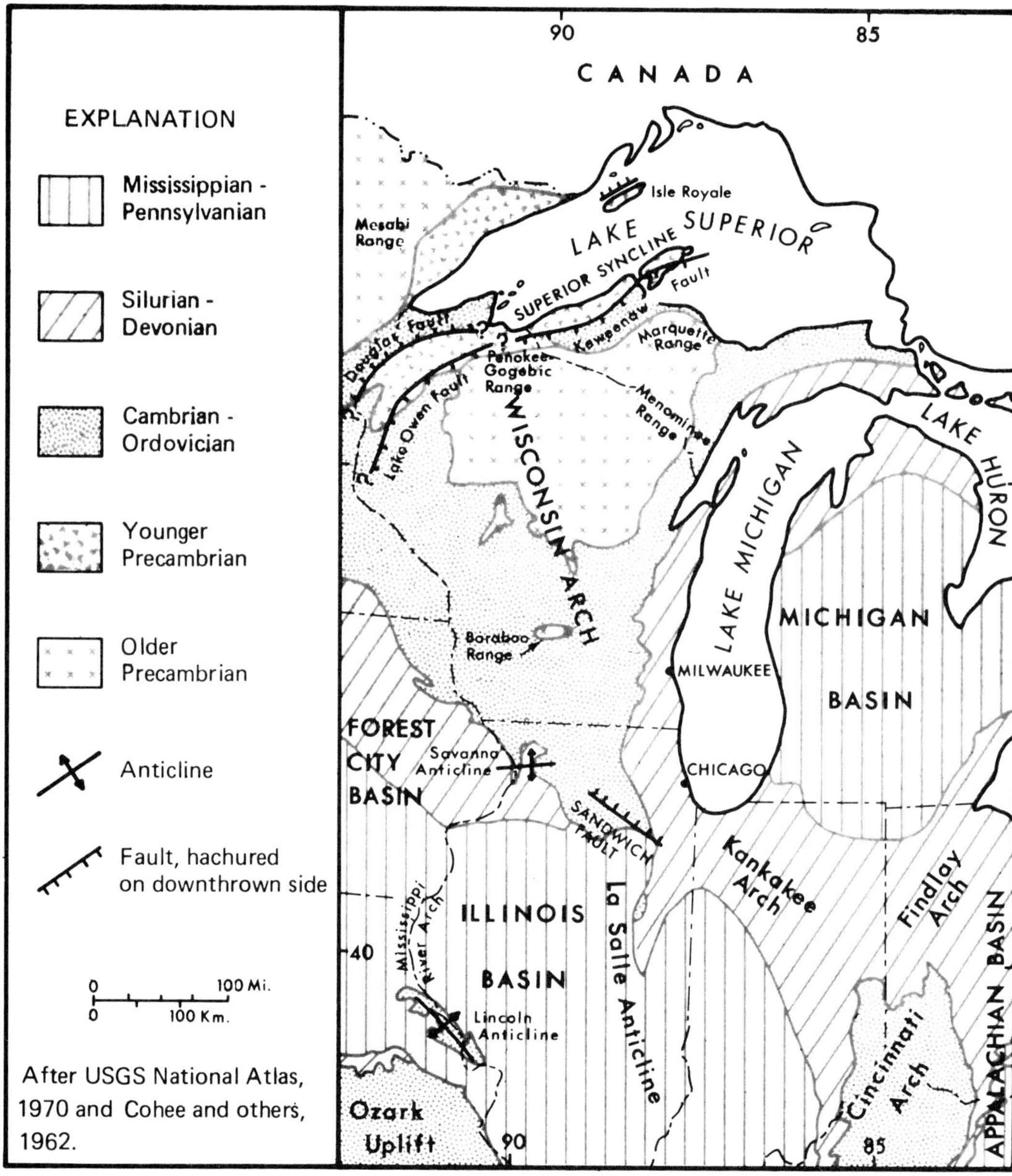

Figure 2-3. Generalized geologic and tectonic map of the north-central United States.

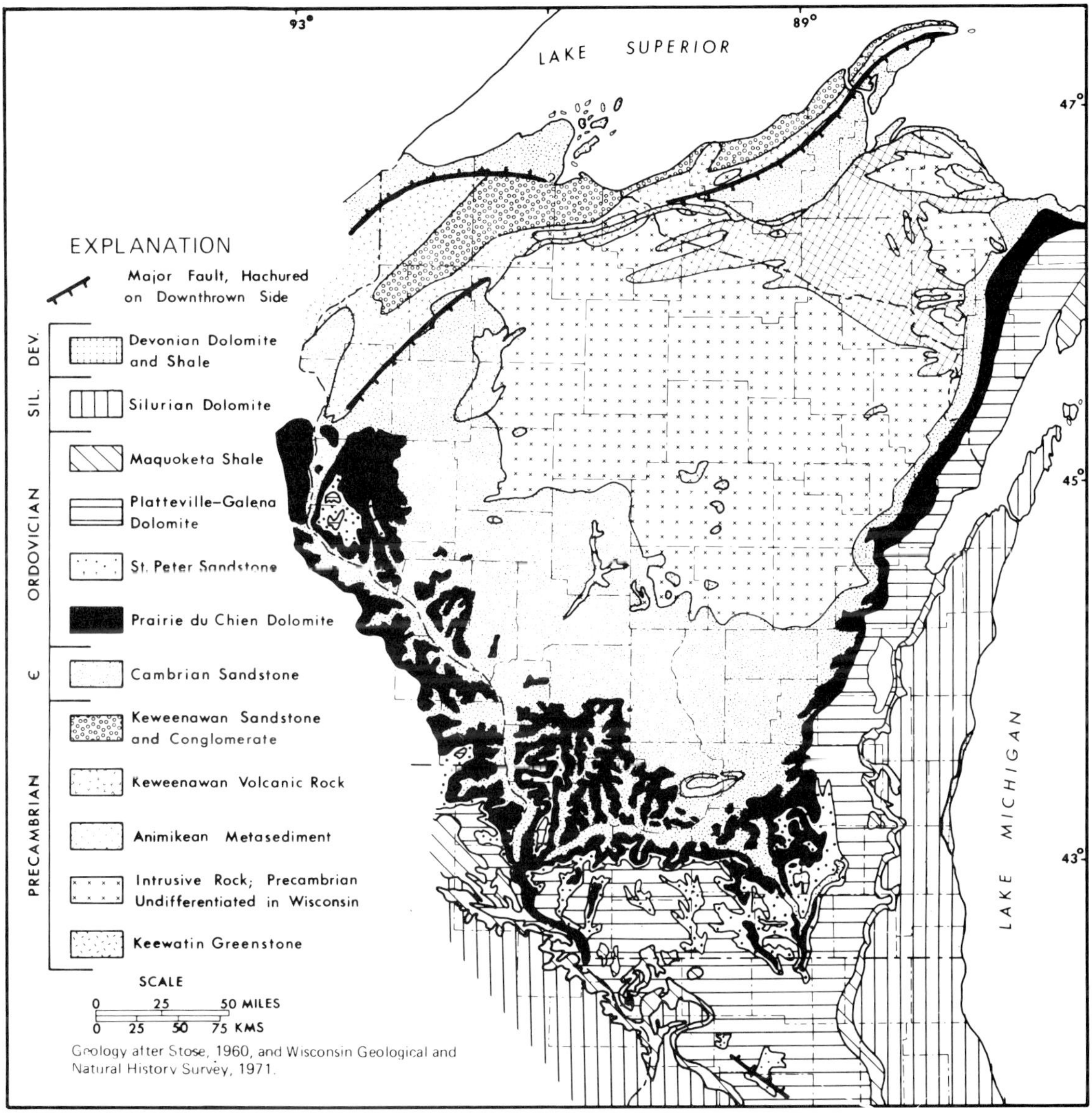

Figure 2-4. Geologic map of Wisconsin, northern Michigan and parts of adjacent states.

Figure 2-5. Pillow structures in metamorphosed Keewatin lavas near Marquette, Michigan. These ellipsoidal bodies were formed when the lavas were extruded under water. A geologic hammer (lower left), which is one foot long, provides the scale.

metamorphosed basic igneous intrusives, lava flows, and shales that owe their dark greenish color to metamorphic minerals such as chlorite and epidote. Some of the more interesting features of these rocks are the characteristic pillow-shaped bodies that form when a lava is extruded underwater (Fig. 2-5). Excellent examples of these pillow lavas can be seen in roadcuts north of U.S. 41 about 5 1/2 miles west of Marquette, Michigan.

The presence of metamorphosed shales within the Keewatin greenstone sequence shows that older rocks were present somewhere in the general area to supply the sediments which formed them. These sediments and the extruded lavas obviously accumulated on an older surface. However, this is all we can say about pre-Keewatin history of this region although older rocks are known from other parts of the Canadian shield.

The Keewatin rocks were intruded by granitic bodies during one or more periods of folding and faulting that may have been associated with mountain building. However, most of the orogenic activity during this time was apparently north of Michigan and Wisconsin. Some 2,400 million years ago the uplifted and complexly folded Keewatin rocks and intruded granites (now largely metamorphosed to gneiss) were leveled by erosion (Fig. 2-7).

Middle Precambrian

A sequence of more than ten thousand feet of Animikean shale and greywacke sandstone, with lesser amounts of iron formation, conglomerate, dolomite, and vol-

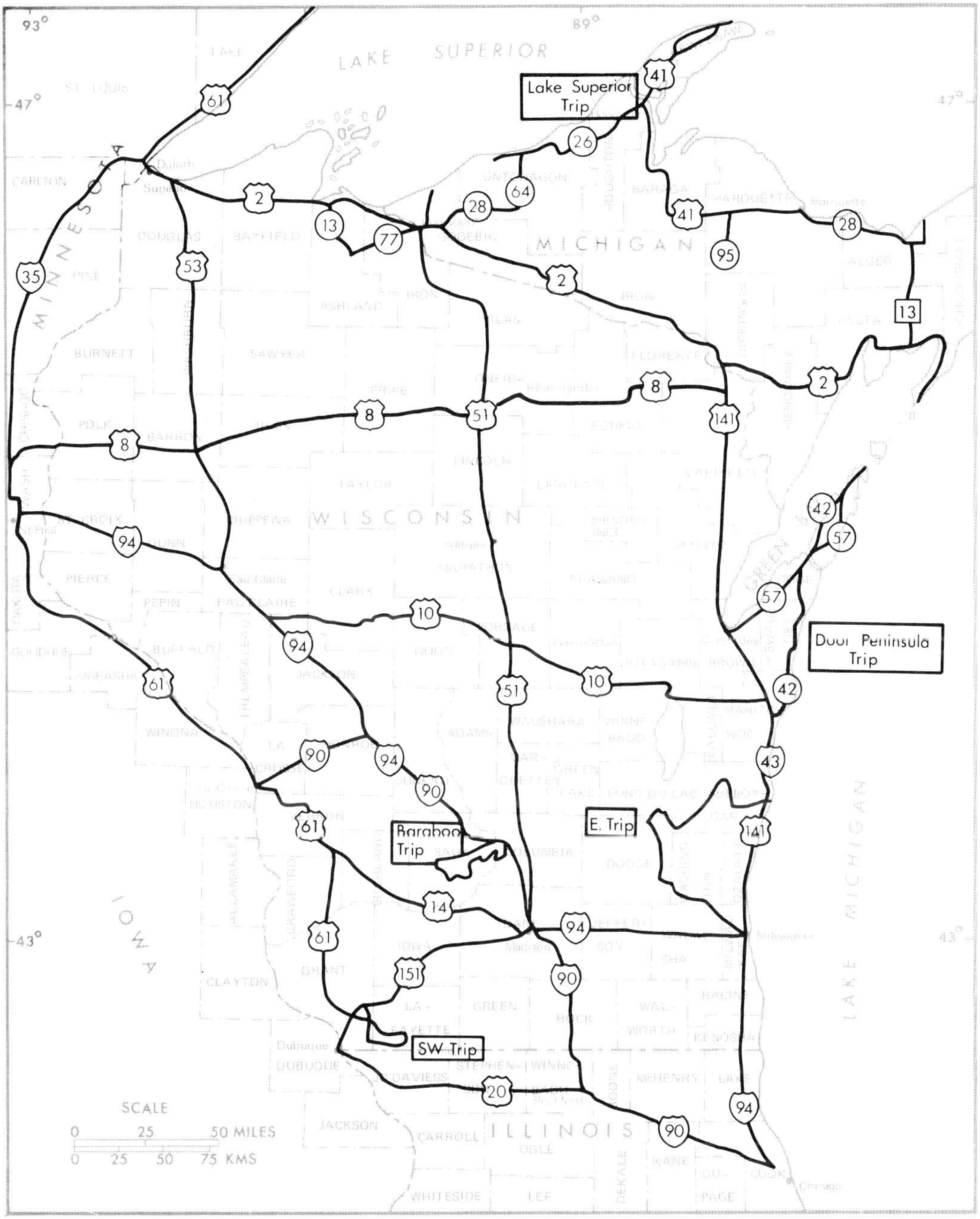

Figure 2-6. County map for Wisconsin, northern Michigan, and parts of adjacent states. Detailed geologic guides are provided in a companion volume for all the highways shown, and for five special trips.

canic rock, was deposited on this newly developed erosion surface as the crust subsided and the area was covered by the sea. Animikean marine deposition was interrupted by periodic uplifts that often were accompanied by igneous activity. Conglomerates overlying unconformities mark these uplifts and define a 3-fold division for Animikean rocks in some areas of northern Michigan (Fig. 2-7).

Animikean rocks are exposed over a large area in northeastern and north-central Wisconsin and in northern Michigan (Fig. 2-4). The iron ores in the old mining districts of Marquette, Penokee-Gogebic, and Menominee are Animikean rocks (Fig. 2-3). These mining areas are called *ranges* because the iron formation and associated quartzite are relatively resistant to erosion and form high-standing linear belts surrounded by more easily eroded Animikean slates.

Iron formation or *taconite* is the most characteristic and unusual rock type in the Animikean sequence. Iron formation is a banded rock composed of silica-rich and iron-rich layers, and it is often reddish in color. In northern Michigan, good examples can be seen at Jasper Knob in downtown Ishpeming and at a roadside park one mile east of Michigamme on U.S. 41. Exposures are also present in all of the old iron mining districts. The iron formation exposed in the taconite mine at Black River Falls, Wisconsin is also an Animikean deposit.

Another interesting Animikean rock type is algal dolomite. This rock establishes that early life was present in the Middle Precambrian seas. Examples of well-preserved algal colonies, that resemble half of a cabbage head in cross section, can be seen up a short trail at the east end of a large roadcut along U.S. 41, about 3 miles southeast of Marquette, Michigan (Fig. 2-8).

Animikean deposition ended some 1,600 million years ago with a widespread mountain building episode that included extensive

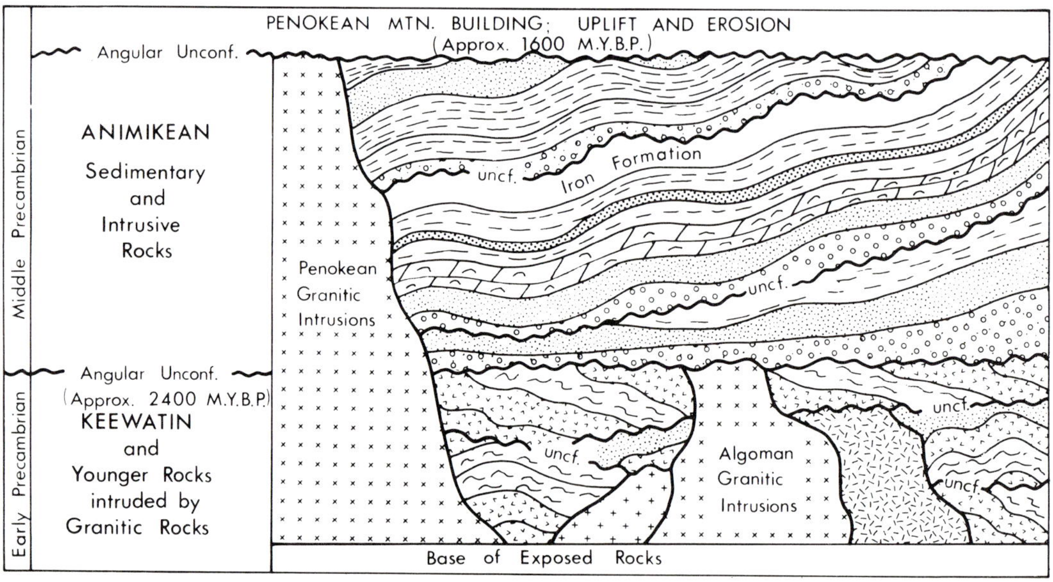

Figure 2-7. Diagrammatic summary of Early and Middle Precambrian geologic history of northern Wisconsin and Michigan. The oldest rocks exposed in this area are of Keewatin age. Lithologic units are generalized from a variety of sources including Dorr and Eschman (1971).

Figure 2-8. Colonies of algae produced these characteristic domed structures in the Middle Precambrian Kona Dolomite near Marquette, Michigan. Numerous small, vertical faults offset the algal laminations at this locality. The hammer provides scale.

igneous intrusion. This event was the Penokean orogeny and the resulting mountains were the Penokean range or Northern Michigan highland. This orogenic belt was apparently south of the older one that involved Keewatin rocks (Fig. 2-9).

The Penokean mountains were formed by north-south compressional forces that resulted in large synclines and anticlines that were often complexly faulted. Metamorphism accompanied the structural deformation and associated igneous intrusions, and slate, schist, quartzite, and marble resulted. This orogeny also involved the previously deformed Keewatin-age units, further complicating the structure and metamorphic character of these rocks.

The Penokean mountains extended across northern Wisconsin and northern Michigan and to the east and west. They must have rivaled, and possibly surpassed, the Alps in magnificent splendor. These remnant hills are still the sites of attractive winter ski resorts, but think of what would have been available in the original Penokean mountains—America's most spectacular mountain playground! This, of course, wasn't to be.

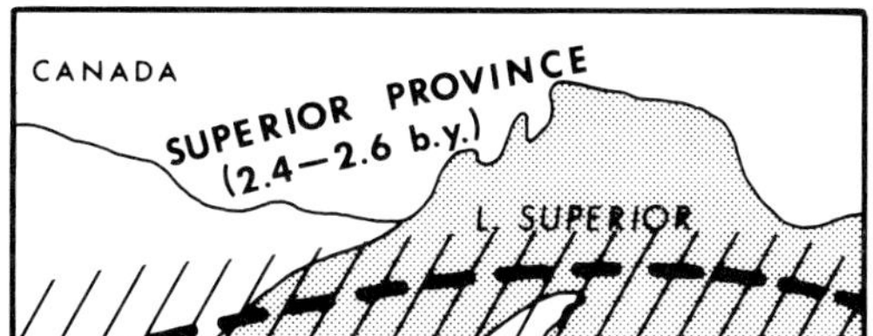

Figure 2-9. Generalized Precambrian orogenic provinces based on absolute age dating of igneous rocks. The general pattern for pre-Keweenawan rocks shows older mountain building episodes to the north toward the main part of the Canadian shield. As indicated by the diagonal lines on the figure, rocks of the Superior Province are present in the Penokean Province of northwestern Wisconsin and northern Michigan. The generalized southern limit of age-dated samples is shown by the dash-dot line. The Wolf River batholith (1.5 b.y.) intrudes older rocks.

Late Precambrian

The majestic Penokean mountains eroded away over a long interval of time as their northern part began to subside into a broad, elongate depression called the Lake Superior syncline. The south edge of this developing depression corresponded to what is now the northwestern part of Wisconsin and the area adjacent to Lake Superior in northern Michigan, while the northern edge was along the Minnesota shore of Lake Superior and on Isle Royale (Fig. 2-10).

The Precambrian history south of the Penokean highlands of Michigan and northern Wisconsin is not well known since exposures are often poor and extensive exploration programs for economic mineral deposits have just begun. A scattering of age-dates from a granitic complex, metamorphosed sedimentary rocks, and volcanics suggests that much of the terrain represents an orogenic province that is somewhat younger than the Penokean orogeny. Thus the pattern of progressively younger Precambrian orogenic belts to the south seems to continue (Fig. 2-9). If these age relations are supported by additional study, the areas of quartzite in Wisconsin at Rib Mountain, the Baraboo Range, and the Barron Hills would be younger than Animikean rocks in Michigan and not involved in the Penokean orogeny, as was assumed for many years. A large, incompletely exposed, granitic intrusive that occupies much of northeastern Wisconsin is dated as early Late Precambrian and it is apparently younger than the quartzites.

Regardless of their ages, these quartzite remnants are all that remain of what once was an extensive cover of Middle to early Late Precambrian sedimentary rocks thousands of feet thick. The folding and subsequent erosion of this sedimentary sequence were accomplished during a long interval that reduced the Precambrian of Wisconsin to a nearly flat plain while deposition was taking place in the Lake Superior syncline to the north.

The subsiding Lake Superior syncline received sands and gravels from erosion of the Penokean highlands in the early Late Precambrian (Lower Keweenawan). However, the Portage Lake lava flows of the Middle Keweenawan are the most impressive deposits in the Lake Superior syncline (Fig. 2-10 and 2-11). The outpourings of these lavas, with some interbedded sandstone and conglomerate, accumulated to more than 20,000 feet before volcanic activity began to wane. What a forbidding but spectacular land it must have been as volcanoes flared and lava flowed in a narrow belt that looped from southern Michigan, across the Upper Peninsula and northwestern Wisconsin, and

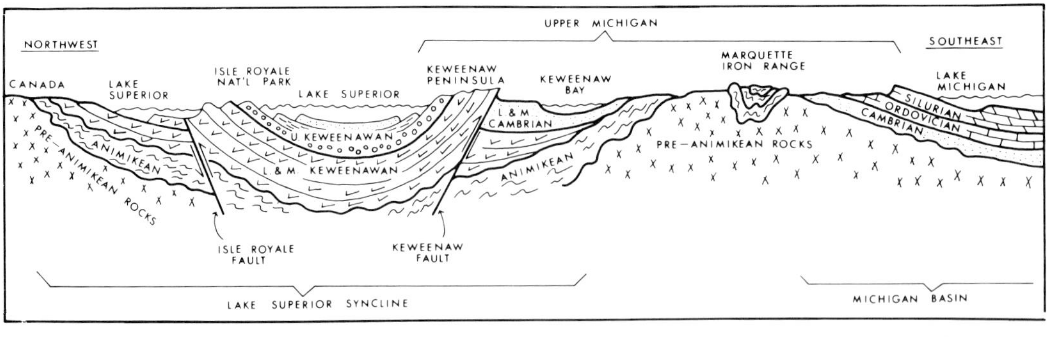

Figure 2-10. Diagrammatic northwest-southeast geologic cross section of the Lake Superior syncline with continuation into the Michigan basin. Major unconformities are illustrated by heavy wavy lines. Lithology and thickness are generalized. The positions of major geologic and geographic features are approximate.

southward across Minnesota all the way to central Kansas.

The relative thinness of individual lava flows (about 100 feet) and the overwhelming cumulative thickness of these outpourings can be appreciated by examining the flows exposed on northern Michigan's Keweenaw Peninsula, along the Minnesota shore of Lake Superior, and on Isle Royale National Park (Fig. 2-12).

Sometime during, or soon after, deposition of the Portage Lake Series, copper mineralization occurred. The copper, often as the native metal, was deposited in vesicles in the lava, between grains in the interbedded sandstone and conglomerate, and within cracks in all rock types. These copper deposits were initially mined by Indians of the Copper Culture and much later they caused the mining boom in the Keweenaw Peninsula and immediately adjacent areas.

About 1,100 million years ago during Late Keweenawan time, volcanic activity diminished and a sequence of conglomerate, shale, and sandstone up to 20,000 feet thick was deposited in the Lake Superior syncline (Fig. 2-10 and 2-11). The basal unit in this Upper Keweenawan sequence is the Copper Harbor Conglomerate. The overlying unit is the copper-bearing Nonesuch Shale that is mined at White Pine, Michigan. The youngest Upper Keweenawan unit, and the final record of the Precambrian, is the thick, feldspathic Freda Sandstone. Much of the sediment deposited in this sequence was derived from the old Penokean area to the south but contributions also came from Lower and Middle Keweenawan rocks that were now exposed along the edges of the depositional basin.

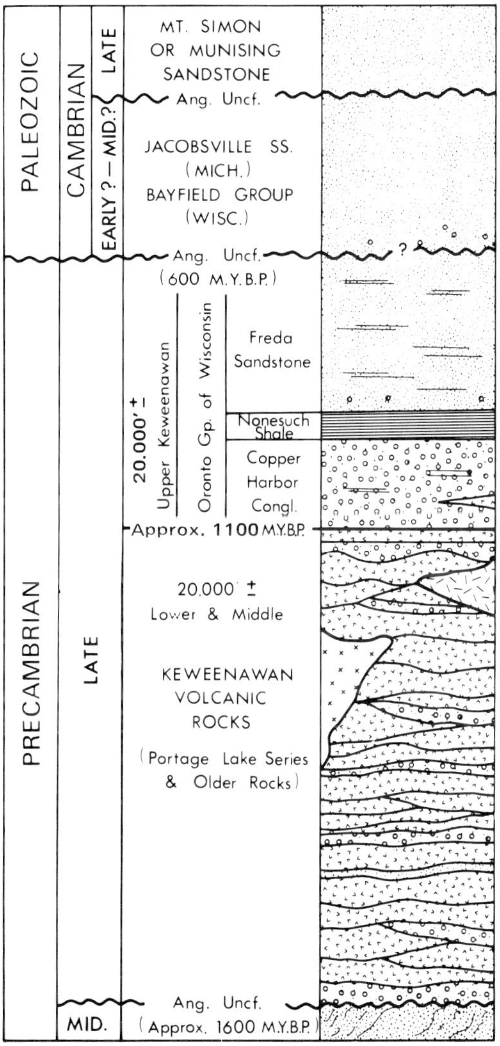

Figure 2-11. Generalized stratigraphic column summarizing Late Precambrian and Early to Middle Cambrian(?) geologic history and rock nomenclature for the Lake Superior syncline. Modified from Dorr and Eschman (1971) and Ostrom (1967).

Latest Precambrian or Early to Middle Cambrian (?)

Unlike the hypothesized dramatic terminations of the Keewatin and Animikean rock sequences, the Precambrian ended in a peaceful manner leaving the Keweenawan rocks essentially undeformed. Apparently there was only minor tilting and uplift of the

Figure 2-12. Middle Keweenawan Portage Lake lavas at the entrance to Rock Harbor on Isle Royale dip southeast toward the center of the Lake Superior syncline.

edges of the Lake Superior syncline before a new depositional sequence of sandstone began.

A succession of unfossiliferous sandstone (Bayfield Group) up to 4,000 feet thick in northwestern Wisconsin and the Apostle Islands may be Lower to Middle Cambrian in age (Fig. 2-11). Similar sandstone units are also present in northern Michigan (Jacobsville Sandstone) and in northeastern Minnesota (Hinckley Sandstone and Fond du Lac Formation). The sands of the Bayfield Group, the Hinckley-Fond du Lac, and the Jacobsville were derived from relatively high-standing land located in the southern part of upper Michigan and in Wisconsin. These sands were apparently deposited in ancient streams and lakes before the invasion of Middle Cambrian seas.

In northern Michigan, the Jacobsville Sandstone is in angular discordance with Animikean and older rocks. This relationship is visible along U.S. 41 at the south end of Keweenaw Bay, and at Presque Isle Point Park immediately north of Marquette, Michigan (Fig. 2-13). The angular unconformity between Fond du Lac and Animikean rocks (Thomson Formation) may be observed along the St. Louis River in Minnesota's Jay Cooke State Park.

Elsewhere in our area, rocks of possible Early to Middle Cambrian age are unknown. Apparently most of this region was undergoing erosion at this time. In northern Michigan, Upper Cambrian rocks are in slight angular discordance with the underlying Jacobsville. This unconformity suggests that the area was slightly uplifted and tilted after Jacobsville deposition and before Late Cambrian time.

Thus we see that the age assignment of the Jacobsville, the Bayfield, and the Hinckley-Fond du Lac to the Paleozoic is largely dependent upon stratigraphic position (Fig. 2-11). There is also some lithologic support for a Paleozoic age assignment. The sand-

Figure 2-13. The irregular unconformity surface separating Early Precambrian gneissic rocks from nearly flat-lying latest Precambrian to Middle Cambrian(?) Jacobsville Sandstone is well defined about halfway up the wave-cut cliff at Presque Isle Park, immediately north of Marquette, Michigan.

stones in these formations are quartz-rich and fairly well sorted. Thus, they resemble Upper Cambrian rocks more than they do the feldspathic sandstones of the Precambrian. However, it is still possible that these rocks are latest Precambrian in age and doubt will remain until age-diagnostic fossils are found.

Lake Superior Syncline

The latest Precambrian to Middle Cambrian (?) sandstones are the last record of deposition in the Lake Superior syncline. This prominent structural feature accumulated up to 40,000 feet of lava, sandstone, and conglomerate in its center (Fig. 2-10). The weight of these rocks, especially the basaltic lava, certainly helped maintain the continued downward movement of this trough. However, the cause of the initial subsidence and the long-term negative tendency of this segment of the crust, during Late Precambrian time and possibly into the Early Paleozoic, is not known.

Associated with the steeply northward dipping, south edge of the Lake Superior syncline is the Keweenaw fault, which extends from Wisconsin through northern Michigan and along the Keweenaw Peninsula (Figs. 2-3 and 2-10). This is a high angle reverse fault with some 4 miles of vertical displacement, and it brings Keweenawan rocks up and outward from the Lake Superior syncline over Jacobsville Sandstone. The fault is expressed along the Keweenaw Peninsula as a prominent escarpment where resistant Keweenawan rocks stand relatively high above lowlands formed on the more easily eroded Jacobsville Sandstone. The fault plane itself is not well exposed but it can be seen at the Natural Wall about 3 miles east of Laurium, Michigan (Fig. 2-14).

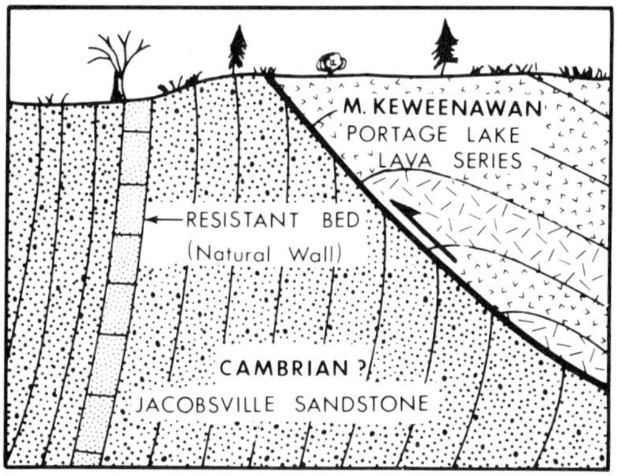

Figure 2-14. Interpretive sketch of an exposure of the Keweenaw fault at Rock Wall Ravine near Laurium, Michigan. This locality derives its name from a near-vertical, unusually resistant, Cambrian(?) Jacobsville Sandstone bed that crops out as a natural wall. The Jacobsville here was dragged up along the fault as Middle Keweenawan Portage Lake volcanics moved up and over these younger rocks.

The Keweenaw fault may extend westward into the Lake Owen fault of Wisconsin (Fig. 2-3). The Lake Owen fault, like the Keweenaw fault, is on the south limb of the Lake Superior syncline and its characteristics are similar. This fault line is marked by a subdued escarpment of Keweenawan volcanics that rises above a low area to the south that is underlain by Cambrian sandstones.

The Douglas fault of Wisconsin is on the north limb of the Lake Superior syncline and it is a south dipping, high angle reverse fault that brings Keweenawan rocks up and over the Bayfield Group (Fig. 2-3). The Douglas fault is well exposed at Amnicon Falls and Pattison Falls state parks near Superior, Wisconsin (Fig. 2-15). Another southward dipping reverse fault on the north edge of the syncline is located north of Isle Royale and it is named for this island (Fig. 2-3).

The ages of the four large faults associated with the Lake Superior syncline are not well established. They apparently started in Late Keweenawan time and movement continued intermittently during deposition of the Lower to Middle Cambrian (?) sandstone sequences. Movement apparently ceased before Late Cambrian time since Upper Cambrian rocks cover the southwestern end of the Lake Owen fault.

The Upper Midwest rightfully takes pride in its stable, relatively earthquake-free geologic environment, but obviously this wasn't always the case. During development of these four major faults, there must have been some spectacular earthquakes that would have tumbled the cities of this region.

PALEOZOIC: THE SEAS COME AND GO

Structural Setting

Depositional patterns of Upper Cambrian and younger Paleozoic rocks in our area reflect the influence of the Wisconsin arch and adjacent basins. The broad, northern part of this arch is called the Wisconsin dome. It was a mildly positive feature that periodically provided sediments to Paleozoic seas which lapped around and over this subdued landmass. The Baraboo Range lies upon the Wisconsin arch and provides a specific, although somewhat unique, example of this general relationship. The Michigan basin, Forest City basin, and Illinois basin flank the arch on the east, west, and south respectively (Fig. 2-3). These basins are negative features that progressively subsided and accumulated a thicker, more continuous succession of Paleozoic sedimentary rocks than the surrounding areas. The location of these basins and the intervening arches is quite

Figure 2-15. The Douglas fault moved resistant Middle Keweenawan lavas up and over flat-lying, less resistant sandstones of the Cambrian(?) Bayfield Group to form this waterfall at Amnicon Falls State Park near Superior, Wisconsin. Wisconsin Department of Natural Resources photo.

well defined by rock distribution patterns on the geologic map (Fig. 2-3), and they also show up well on cross sections (Fig. 2-16).

As previously described, the area of the Wisconsin dome was emergent and contributed sediments to the subsiding Lake Superior syncline in Keweenawan and possibly earliest Paleozoic time. This area was part of a vast region of eroding Precambrian rocks that included much of the North American continent. In the Late Cambrian the seas transgressed over this ancient land surface.

During Late Cambrian and Early Ordovician deposition, a faint suggestion of a Wisconsin arch developed when an embayment formed in the area of western Wisconsin, eastern Iowa, and southeastern Minnesota. The well defined, positive Transcontinental arch that projected southward from the Canadian shield into northern and western Minnesota formed the western edge of this embayment. Another Late Cambrian embayment formed to the south and east, and extended northeasterly through the areas that would later become the Illinois and Michigan basins. This embayment may have continued as far north as the Munising area of northern Michigan where a thick sequence of Late Cambrian sandstones accumulated.

Although these two embayments defined

an arch, it was intermittently covered during Late Cambrian and Early Ordovician time. Late in Early Ordovician time the Wisconsin arch and its southeasterly extension, the Kankakee arch, were emergent and subjected to intense erosion (Fig. 2-3). Movement along the Sandwich fault, that trends southeast through Lee and Ogle counties in Illinois, was apparently started by the development of the Kankakee arch. Lower Ordovician and some Upper Cambrian rocks were eroded from the exposed arches prior to Middle Ordovician deposition.

When the Middle Ordovician seas returned, the Wisconsin arch was well defined and the Illinois and Michigan basins were separated by the Kankakee arch. A broad, poorly defined, largely submergent feature called the Mississippi River arch also provided some definition for the western edge of the Illinois basin at this time (Fig. 2-3). To the west of this arch was the subtle depression that would later develop into the Forest City basin (Fig. 2-3). This depression persisted northward into southeastern Minnesota as an embayment between the Wisconsin arch on the east and the Transcontinental arch on the west.[2]

Although the arches that separate basins of the Upper Midwest were formed by Late Ordovician time, they were no longer emergent features, and the Late Ordovician and Silurian seas covered them. However, the water depth over the Kankakee and Wisconsin arches was shallow enough to allow Silurian reefs to develop around the edges of the Michigan basin.

In Late Silurian and Early Devonian time, the arches and parts of adjacent areas were uplifted again and subjected to erosion. The seas advanced over the arches by the Late Devonian and deposition continued into the Early Mississippian.

In Late Mississippian and Early Pennsylvanian time, northern Michigan, Wisconsin, much of Minnesota, northeastern Iowa, and Illinois north of the Kankakee arch were gradually uplifted and subjected to erosion. The Michigan, Illinois, and Forest City basins remained as negative areas that received sediments until late in the Pennsylvanian. By Permian time the role of arches and basins in the depositional history of this area was over, the entire region was uplifted and a long erosional interval began.

South of the Kankakee arch, a major structural trend also developed during this time (Fig. 2-3). This trend includes the Savanna anticline of easternmost Iowa and northwestern Illinois. The Savanna can be traced eastward into the southerly trending LaSalle anticlinal system which projects into the Illinois basin. Formation of the LaSalle anticline apparently caused the south side of the nearby Sandwich fault to move upward hundreds of feet. Once again the Upper Midwest probably had some dandy earthquakes.

Several other tectonic features are shown in Figure 2-3 but they play a less important role in the historical geology of our area. The Findlay arch trends northeasterly from the Kankakee arch to join the prominent Cincinnati arch. The Wisconsin, Kankakee and Findlay arches, along with the Precambrian shield to the north, define the edges of the Michigan basin. East of the Findlay and Cincinnati arches is the Appalachian basin, another negative feature which accumulated a thick succession of Paleozoic rocks.

The limits of the Illinois basin are also shown on Figure 2-3. This basin is broadly defined by the Kankakee arch on the northeast, the Cincinnati arch on the east, the Ozark uplift on the southwest, and the Lincoln anticline and Mississippi River arch on

2. This feature is known as the Hollandale embayment.

the west. The latter two features and the Ozark uplift also define the eastern edge of the Forest City basin. Thus, we had many large basins separated by relatively narrow arches in the stable interior of North America during the Paleozoic.

As described above, the development of the Wisconsin, Kankakee, and Mississippi River arches was not a process that involved the uplift of positive features. These arches were merely less subsident areas along the margins of the more rapidly subsiding Michigan, Illinois, and Forest City basins. The net effect of this relationship can be appreciated when we realize that Precambrian rocks exposed at the surface in northern Wisconsin are overlain by more than 14,000 feet of Paleozoic rock in the approximate center of the Michigan basin in southern Michigan (Fig. 2-3 and 2-16). However, at no time during deposition did an elevation differential anywhere near this magnitude exist between the Wisconsin arch and the Michigan basin. In most instances, the sedimentary rocks that accumulated in the basins were deposited in shallow seas. Although individual formations are much thicker in the basins, they are lithologically similar to correlatives that accumulated on the arches. The sedimentary section on the

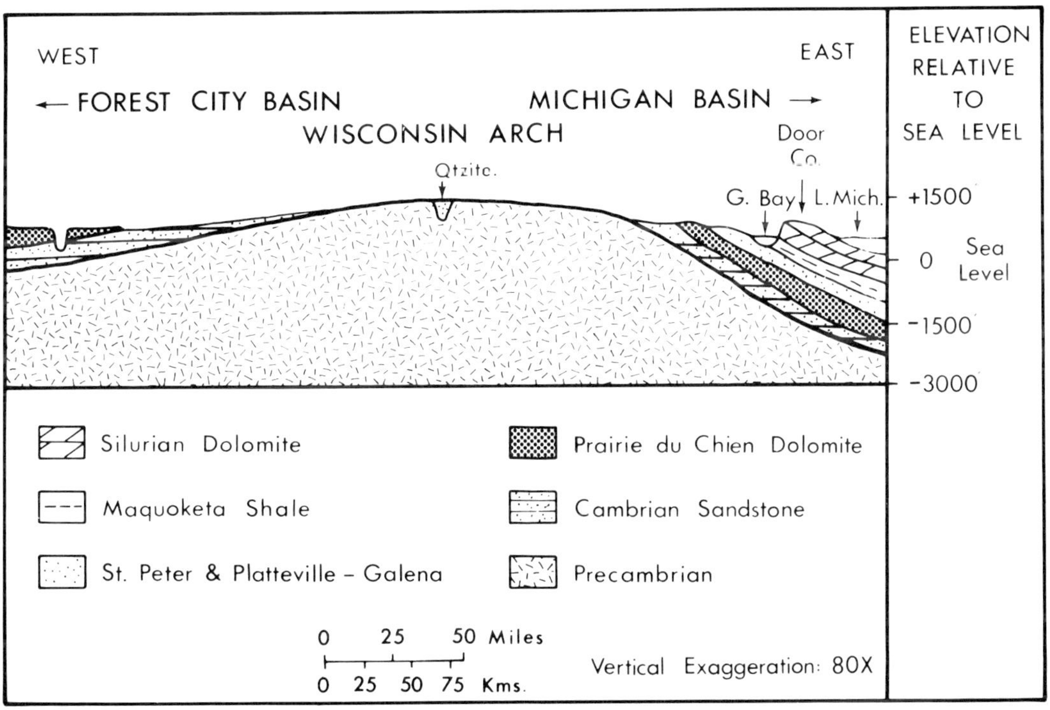

Figure 2-16. Generalized east-west geologic cross section along 45 degrees north latitude across Wisconsin to easternmost Minnesota (see Figure 2-4 for location of this cross section). The difference in the vertical and horizontal scales (80 times) exaggerates the prominence of the Wisconsin arch and the dip of the Paleozoic rocks off the arch into adjacent basins on the east and west. The dip into the Michigan basin is actually about 3 to 5 degrees.

arches was also thinned by erosion in late Early Ordovician, Late Silurian to Early Devonian, and during latest Mississippian to Early Pennsylvanian while the arches were emergent.

Late Cambrian

Our Upper Cambrian rocks are sandstone, shaly sandstone, dolomite, and sandy dolomite. The basal Upper Cambrian deposit is a well-sorted, medium-grained, quartz sandstone that was deposited by a sea transgressing over igneous or metamorphic Precambrian rocks. This unconformity is exposed in numerous places, but it is especially accessible at Irvine Park in Chippewa Falls, Wisconsin, and below the dam at Nekoosa, Wisconsin.

The basal Upper Cambrian deposit is called the Mt. Simon Sandstone for Mt. Simon, a hill at Eau Claire, Wisconsin.[3] The sand in this formation represents extensively weathered material derived from Precambrian rocks. After prolonged weathering, only the most resistant minerals, primarily quartz, were available for transportation into the encroaching seas. Once the sand reached the sea, it was distributed by waves and currents to form a relatively thick (300-800 feet), blanketlike deposit with ripple marks, cross-bedding, and worm borings attesting to a shallow water heritage.

The depositional edge of the Mt. Simon did not reach northern Michigan, which was a highland area. However, eroding Jacobsville on the northern side of this highland contributed sediment to the north where a sandstone that is time equivalent to the Mt. Simon was deposited. This is the Chapel Rock Sandstone and it is the basal member of the Upper Cambrian Munising Formation.

As Cambrian seas continued to encroach onto the Precambrian shield, waters deepened offshore and depositional rates slowed as the source area for sand shifted further north. This resulted in the deposition of a clayey, often shaly-bedded, sometimes dolomitic, finer grained sandstone called the Eau Claire Formation. The Eau Claire is about 250 feet thick and contains abundant sand-size, green glauconite (a potassium-iron-magnesium aluminum silicate mineral). Sands rich in glauconite are commonly called greensands, and it is believed that they accumulated quite slowly.

Fossils such as phosphatic brachiopods and trilobite fragments are fairly common in deeper, quieter water deposits such as the Eau Claire, while they are rare in cleaner, coarser, quartz sandstone units such as the underlying Mt. Simon.

A relatively rapid regression of the sea after deposition of the Eau Claire resulted in erosion of the top of this formation prior to deposition of the overlying coarser, better-sorted, quartz sandstone of the Wonewoc Formation. The Wonewoc is another thick (100 to 400 feet), transgressive, shallow-water sandstone much like the Mt. Simon, and it marks the beginning of a new cycle of sedimentation. Wonewoc exposures contribute much to the scenery of central Wisconsin. Such features as the Dells of the Wisconsin River, Mill Bluff, and Roche A Cri Mound are composed of this sandstone (Fig. 2-17). Overlying the Wonewoc is the Tunnel City (or Franconia) Sandstone, which is 100 to 200 feet thick, and lithologically similar to the Eau Claire. This suggests a continuation of the transgression recorded by the Wonewoc, with the Tunnel City or Franconia representing a deeper water de-

3. In describing Paleozoic history we haven't used all the available formational names nor do we detail all events. For additional data on rock nomenclature the reader is referred to Ostrom's (1967) Information Circular 8 of the Wisconsin Geological and Natural History Survey.

Figure 2-17. Scenic, water-carved exposures of cross-bedded Middle Cambrian Wonewoc Sandstone line the Lower Dells of the Wisconsin River in Sauk County, Wisconsin.

posit further removed from the source of sediments and with slower rates of sedimentation.

A limestone was the uppermost deposit in the transgressive cycle when water depth and restricted supplies of sand or mud permitted. These limestones were usually converted to dolomite soon after deposition. Such a dolomite overlies the Tunnel City (or Franconia) as the uppermost unit in this cycle. This is the St. Lawrence Formation and it is 50 to 125 feet thick. A distinctive, shaly-bedded, dolomitic siltstone called the Lodi Member is present in the upper St. Lawrence. The Lodi contains well-preserved trilobite fragments, phosphatic brachiopods, and merostomes at several localities in Wisconsin. After deposition of the Lodi, the sea once again retreated and the St. Lawrence was locally removed by erosion.

This repetition of rock sequences, consisting of a basal massive-bedded quartz sandstone that is unconformable on older rocks and gradational into younger units of clayey, often glauconitic or dolomitic, sandstone which may upgrade into carbonate, occurred five times during the Late Cambrian and Ordovician in this area (Table V). This cyclic pattern records repeated transgressions and regressions of the sea out of adjacent basins and onto the developing Wisconsin arch.

In the shallower waters on the crest of the arch, wave and current action were vigorous, and offshore sandstones such as the Eau Claire and Tunnel City (or Franconia) could not develop. Under these conditions, carbonate deposition was usually impossible, and the cycles described above may not be recognizable.

In Wisconsin, northern Illinois, eastern Iowa and eastern Minnesota, each major lithologic change in this cyclic sequence has been named. In northern Michigan, however, only one depositional cycle is present. The Upper Cambrian part of this cycle is defined by the basal Chapel Rock Sandstone and the overlying Miner's Castle Sandstone. These units are the two members of the Munising Formation, which is spectacularly exposed at Pictured Rocks National Lakeshore east of Munising, Michigan.

As previously mentioned, the Chapel Rock was deposited as a nearshore sandstone north of a highland that existed in upper Michigan early in Late Cambrian time. The upper member of the Munising is a transgressive high energy sand deposit that largely covered the old highland area. The Cambrian-Precambrian unconformity at the old Quinnesec iron mine along U.S. 2 at Quinnesec, Michigan is a record of this transgression. Here, pebbles from the folded Precambrian Vulcan Iron Formation are incorporated into flat-lying, upper Munising Sandstone to form a spectacular basal conglomerate.

The Jordan Sandstone is the youngest Cambrian deposit in our area and it is a clean, well-sorted, white, medium-grained sandstone about 20 to 150 feet thick. This is another high energy sandstone that was deposited in a shallow sea transgressing over an unconformity onto the arch. As such, it marks the base of our third depositional cycle.

Early Ordovician

As the Jordan transgression continued, the sea deepened and some carbonate and shaly-bedded sandstone were deposited at the top of the Jordan. These conditions persisted, and the Jordan graded into the overlying Oneota Dolomite of Early Ordovician age. Since this lithologic change is a continuation of a transgressive cycle, no unconformity separates Cambrian from Ordovician rocks in this area.

It would be helpful if all the geologic periods were defined at the top and bottom by recognizable unconformities. Sometimes this is the case because of the way in which geologic periods were established. In the type areas where the Early and Middle Paleozoic periods were originally established, they are separated by unconformities. Away from these areas the unconformities may not be present. We see a similar situation in our own area, where unconformities present on the Wisconsin arch lose definition in adjacent basins where the seas continued to receive a sedimentary record.

We noted above that the Lower Ordovician Oneota Dolomite represents the deeper part of the transgressive cycle initiated during deposition of the Late Cambrian Jordan Sandstone. However, the Oneota was not deposited in water that was deeper than 100 feet. Furthermore, ripple marks, mud cracks, and undulating, thin laminations formed by algae suggest much of the Oneota was deposited in the tidal zone (Fig. 2-18). Fossils are poorly preserved in this formation because of dolomitization of the original limy shell materials. Fossils are also rare

TABLE V
Summary of Paleozoic History

Stratigraphic Position	Depositional Situation and Cycle No.	Formation and Thickness in Ft.	
LOWER MISSISSIPPIAN	No record, but possible deposition of shales and limestones followed by uplift and erosion in Late Mississippian time		
MIDDLE AND UPPER DEVONIAN	Transgression and carbonate deposition	Devonian dolomite and limestone, 230'±	
	Regression ----------	Unconformity ----------	
SILURIAN	Transgression and carbonate deposition	Silurian dolomites, 400'-600'	
	Regression ----------	Unconformity ----------	
UPPER ORDOVICIAN	Transgression and carbonate deposition? or Regression	Neda Formation, 0-55'	
	Regression?	Unconformity?	
	Offshore; orogenic muds	Maquoketa Shale, 100'-350'	
----------	Regression ----------	Unconformity ----------	
MIDDLE ORDOVICIAN	Carbonate deposition	Platteville-Decorah-Galena Fms., 200'-350'	
	Offshore CYCLE V	Glenwood Formation, 20'±	
	Transgression	St. Peter Sandstone, 40'-350'	
----------	MAJOR REGRESSION ----------	Unconformity ----------	
LOWER ORDOVICIAN (Prairie du Chien Dolomite, 150'-350')	Carbonate deposition	Willow River Dolomite, 30'-60'	
	Transgression CYCLE IV	New Richmond Sandstone, 5'-50'	
	Regression ----------	Unconformity ----------	
	Carbonate deposition	Oneota Dolomite, 100'-200'	
	CYCLE III		
UPPER CAMBRIAN	Transgression	Jordan Sandstone, 20'-150'	
	Regression ----------	Unconformity ----------	Munising Fm. (U. Mich.)
	Carbonate deposition	St. Lawrence Formation, 50'-125'	
	Offshore CYCLE II	Franconia or Tunnel City Sandstone, 100'-200'	
	Transgression	Wonewoc Sandstone, 100'-400'	
	Regression ----------	Unconformity ----------	
	Offshore	Eau Claire Sandstone, 250'±	
	CYCLE I		
	Transgression	Mt. Simon Sandstone, 300'-800'	
----------	Angular Unconformity	----------	
LOWER-MIDDLE CAMBRIAN(?)	Jacobsville Sandstone (Mich.) or Bayfield Group (Wisc.) Hinckley and Fond du Lac Sandstones (Minn.)		
----------	UNCONFORMABLE WITH UNDERLYING PRECAMBRIAN	----------	

Figure 2-18. Small colonies of algae formed these biscuitlike structures on the bedding surface of a slab of Lower Ordovician Prairie du Chien Dolomite; Buffalo County, Wisconsin. The hammer provides scale.

and this suggests an unfavorable environment for animal life during deposition of the Oneota.

In northern Michigan, the Oneota equivalent may be the Au Train Formation. This relatively thin dolomitic sandstone unit forms the top of the cycle initiated in the Late Cambrian by deposition of the Munising Formation. The Au Train forms a cap rock that protects the less resistant Upper Cambrian sandstones. This cap rock holds a prominent escarpment east from Marquette, Michigan, and a series of waterfalls (Laughing Whitefish, Au Train, Munising, Miner's, etc.) are developed along it (Fig. 2-19).

After deposition of the Oneota the seas again retreated for a brief period of time, and a minor erosional unconformity developed. A new Early Ordovician transgression began and the Shakopee Formation was deposited. Thc Shakopee has a basal transgressive sandstone member (New Richmond Sandstone) and an upper dolomite member (Willow River Dolomite) that is lithologically like the Oneota. These two members of the Shakopee comprise our fourth transgressive cycle. In spite of the unconformity between the Oneota and the Shakopee, these two formations are conveniently grouped as the Prairie du Chien Dolomite (or Group) which is 150-300 feet thick (Fig. 2-4).

The Prairie du Chien Dolomite, like the Au Train of Upper Michigan, is a unit that provides a cap rock for the less resistant underlying Cambrian rocks. The uplands of southwestern and western Wisconsin, eastern Minnesota and eastern Iowa are capped with Prairie du Chien rocks.

Middle Ordovician

At the end of Early Ordovician time a regional uplift occurred, the seas withdrew once again and an intensive interval of erosion began. Recently deposited rocks were eroded throughout the region, and more than 500 feet of rock were removed in some localities to expose units as old as the Eau Claire. The arches were affected most severely but parts of the adjacent basin areas were also eroded. A new transgression is indicated by the basal conglomerate of the St. Peter Sandstone which includes materials reworked from underlying rocks. This basal deposit is followed by the distinctive white, massive-bedded, medium-grained, well-sorted quartz sandstone that sometimes weathers reddish (Fig. 2-20). Many of the well-rounded quartz grains have wind-frosted surfaces, indicating that part of the St. Peter may have been deposited as dunes. Cross-bedding is also a common feature.

An erosional channel cut into the Prairie du Chien surface that is filled with St. Peter Sandstone is exposed along Wisconsin Highway 27, 1.5 miles east of the intersection of Wisconsin 35, 60 and 27 in Prairie du Chien. Another St. Peter-filled erosional

Figure 2-19. Miner's Falls in Pictured Rocks National Lakeshore, Alger County, Michigan, plunges over an escarpment of resistant Lower Ordovician Au Train Formation. This formation protects the softer, more massive appearing, underlying Upper Cambrian Munising Sandstone from erosion. To the right of the falls and behind the falling tree, thin, even, resistant beds of the Au Train overlie the sandstone. This contact approximately defines the Cambrian-Ordovician contact. Michigan Tourist Council photo.

Figure 2-20. An unusually resistant cliff of Middle Ordovician St. Peter Sandstone forms Enee Point, near Cox Hollow Lake in Governor Dodge State Park, Iowa County, Wisconsin. Wisconsin Department of Natural Resources photo.

channel is exposed on County G about 0.25 mile west of Mt. Vernon (southwest of Madison about 12 miles); here the St. Peter is unconformable on the Oneota. Since the St. Peter was deposited upon such an irregular erosion surface, its thickness varies from 40 to 350 feet.

The sandstone of the St. Peter grades upward into the thin (20 feet), dolomitic, shaly-bedded sandstone and shale of the Glenwood Formation. The Glenwood reflects deeper water as the Middle Ordovician transgression continued. Apparently this transgression did not reach northern Michigan during St. Peter-Glenwood time since rocks of this age are not found there (Fig. 2-4). However, the younger carbonate part of the transgressive sequence is present.

A continuation of this transgression resulted in deposition of about 200-350 feet of Middle Ordovician Platteville-Galena dolomites (Fig. 2-4). This sequence consists of massive- to thin-bedded, sometimes sandy, dolomitic limestone (Platteville Dolomite), with a shaly bedded interval in the middle (Decorah Formation), and an upper massive dolomite unit (Galena Dolomite). This rock succession indicates normal marine, moderately shallow water deposition that was interrupted during the accumulation of the Decorah by an influx of muds from the Transcontinental arch to the west.

This sequence is one of the more fossiliferous units in the area. Brachiopods, gastropods, and bryozoans are abundant, so collecting can be a rewarding experience. The unique, spongelike *Receptaculites* is fairly common in the upper part (Galena Dolomite) of this sequence and it is a good guide fossil. In northern Michigan, equivalent rocks are the fossiliferous Black River and Trenton Groups.

Late Ordovician

In Late Ordovician time a minor unconformity developed in parts of this region. However, conformable relations exist in northern Michigan, Minnesota, and northeastern Iowa. When the seas returned, they were transporting muds derived from the Taconic orogeny that was taking place in the northern Appalachian Mountains. Deposits formed in this muddy sea are now the Maquoketa Shale (called Richmond in northern Michigan). The change in sediment supply that resulted in deposition of the Maquoketa breaks the pattern of cyclic sedimentation that existed in this area from the beginning of the Late Cambrian.

The Maquoketa is greenish shale with interbedded nodular limestone and thin-bedded dolomitic limestone. It varies in thickness from 100 to 350 feet. Thickness and lithology of the Maquoketa reflect the influence of the eastern source area. The Maquoketa is thickest (about 350 feet) in northeastern Wisconsin, while it thins to about 100 feet in the southwest part of our area. In eastern Iowa, it contains less clayey material and more carbonate (Fig. 2-4).

Good exposures of the Maquoketa are rare since it is a soft, easily eroded deposit. However, there are outcroppings of this formation in eastern Iowa, near Mt. Carroll, Illinois; at Wisconsin's Blue Mound State Park, and at a Calumet County park just north of Stockbridge, Wisconsin. It is also exposed in a quarry just south of Oakfield, Wisconsin (Fond du Lac County) and at a roadside park about 10 miles north of Green Bay, along Wisconsin 57 (Fig. 2-21). Fossils are abundant at all these localities.

The Upper Ordovician, or Lower Silurian, Neda Formation locally overlies the Maquoketa. The Neda is an unusual sedimentary iron ore that may be as thick as 55 feet. This formation was mined between Iron Ridge and Mayville, Wisconsin during the late 1800s and early 1900s. The ore contained phosphate which makes steel brittle, so the mines eventually closed.

Figure 2-21. Resistant ledges of Silurian dolomite gradationally overlie less resistant, thinner-bedded, Upper Ordovician Maquoketa Shale at Wequiock Falls Roadside Park near Green Bay, Wisconsin.

Originally the Neda consisted of small (less than "BB" size), rounded pellets composed of thin concentric layers of calcium carbonate which formed around a sand grain or fossil-fragment nucleus. These pellets are called *oolites* and they form today in warm, shallow seas where waves and currents are active. After formation, and before deposition of the overlying Silurian dolomite, the Neda oolites were converted by uncertain chemical processes to oolites of iron oxides and silicates. Even the fossils in the formation were converted to iron compounds.

The thicker Neda deposits of eastern Wisconsin seem to have formed as oolitic sandbars similar to those found in modern seas. These bars probably formed during regression of the sea at the end of the Ordovician. Storm waves and currents spread the oolites over a wide area and they are locally present as thin layers in eastern Iowa and northern Illinois. This formation is exposed at Neda, Wisconsin (Dodge County); below Kittell Falls about 4 miles east of DePere, Wisconsin on County G; near Dubuque, Iowa, and in Kankakee River State Park, Illinois.

The age of the Neda is uncertain. The few fossils collected from this unit suggest it is latest Ordovician. However, some geologists believe the Neda is unconformable on the underlying Maquoketa and these fossils were reworked from this unit during erosion. If so, this could mean that the Neda is Lower Silurian. We believe this formation is gradational with the Maquoketa and of latest Ordovician age.

The Neda is unconformable with the overlying Silurian (Fig. 2-22). Where the Neda is not present in eastern Wisconsin and northern Michigan, the Maquoketa (Richmond in northern Michigan) seems to grade into overlying Silurian dolomites. In eastern Iowa and northern Illinois, however, there is a well developed unconformity at the top of the Maquoketa.

Figure 2-22. Silurian dolomite in unconformable contact with the less resistant, underlying Upper Ordovician Neda Formation at an abandoned open-pit iron mine near Mayville, Wisconsin. The lowest beds of dolomite thin and disappear when traced from right to left across the photo. This suggests that the Neda at this locality was a moundlike feature which was gradually covered by Silurian deposits.

Silurian

A thick (400 to 600 feet) sequence of Lower to Upper Silurian dolomites overlies the Maquoketa or Neda. These rocks were deposited during a major marine transgression and they are well-bedded, dense, light-colored dolomite that may contain chert nodules and few marine fossils. In northern Illinois and eastern Wisconsin massive reef deposits are locally present. Examples of coral reefs are found in the vicinity of Thornton, Illinois, and near Racine and Reedsville, Wisconsin. The most striking aspects of these reefs are the steeply dipping reef-flank beds composed of material broken off the growing reef by waves.

The reefs of eastern Wisconsin and northeastern Illinois are part of a system that surrounded the shallow margins of the Michigan basin during Silurian time. The growth of these reefs restricted the circulation of the sea within the Michigan basin and the salt concentration increased so that several thousand feet of gypsum and rock salt (halite) were deposited in the central part of the basin.

Outward from the reef zone, shallow seas free from land-derived materials accumulated thick sequences of fine limy sediments that became well-bedded dolomites. These conditions were in marked contrast to the muddy Late Ordovician seas.

The Silurian dolomites are resistant units and form excellent outcrops. One of the most impressive of these outcrop areas encircles the northern part of the Michigan basin all the way from Rochester, New York through the Garden Peninsula of northern Michigan to Wisconsin's Door County Peninsula, and along the east edge of Lake Winnebago to Milwaukee (Fig. 2-23). The rocks

Figure 2-23. Silurian dolomite along the shore of Lake Michigan at Cave Point, Door County, Wisconsin. Wisconsin Department of Natural Resources photo.

along this trend dip into the Michigan basin at 3 to 5 degrees (Fig. 2-16). This feature is called the Niagaran escarpment or cuesta, since Niagara Falls flows over its prominent edge in New York.

The resistant Silurian dolomites also cap several high isolated hills in southwestern Wisconsin. These include Blue Mound, Belmont Mound, and Platte Mound. These features are outliers from more continuous exposures in northwestern Illinois and eastern Iowa (Fig. 2-4). The outliers help to establish that Silurian rocks once were more widespread than their present distribution suggests. In fact, the Silurian seas covered our entire area.

Devonian

At the end of the Silurian, much of the region was gently uplifted and subjected to erosion. The overlying basal Devonian rocks in eastern Iowa (just off the map edge on Fig. 2-4), northern Illinois (south of the map edge on Fig. 2-4), and in the Milwaukee area are of Middle Devonian age. Thus, this unconformity represents all of the Early Devonian. However, Lower Devonian rocks are known from drilling in the deeper parts of the Michigan and Illinois basins.

Middle to Upper Devonian rocks in the Milwaukee area are about 230 feet thick. The succession consists of medium-bedded dolomite at the base, argillaceous limestone in the middle, and black shale at the top. This black shale unit (Kenwood Formation) is known from excavations under Lake Michigan just offshore from Milwaukee. Devonian outcrops in Wisconsin are scarce. However, fossiliferous outcrops at Estabrook Park in Milwaukee and at Wisconsin's Harrington Beach State Park in Ozaukee County are available for study (Fig. 2-24).

The Devonian rocks of eastern Iowa, northern Illinois, and eastern Wisconsin are primarily carbonate rocks with no indication that they were shoreline deposits. This suggests that Middle and Upper Devonian rocks originally covered much of our area but they were subsequently removed during the long erosional interval that followed a Late Mississippian uplift.

Mississippian

Although we have no positive evidence of Mississippian rocks in our area, it is probable that Lower Mississippian shales were deposited over at least part of this region. This belief is based on the presence of Lower Mississippian rocks in all of the adjacent basins (Fig. 2-3). Like the Devonian rocks, the eroded edges of these Mississippian formations are not shoreline deposits.

After the Early Mississippian, much of this region was subjected to a major uplift. The seas withdrew and an erosional interval that lasted until the Pleistocene was initiated. Paleozoic seas lingered, however, in the deeper parts of the Forest City, Illinois, and Michigan basins and deposited thick sequences of Missisippian and Pennsylvanian rocks. The seas withdrew from these basins prior to the Permian and the marine sedimentary history of this area ended.

Figure 2-24. Middle Devonian dolomite and dolomitic shale exposed along the Milwaukee River at Estabrook Park, Milwaukee, Wisconsin.

MESOZOIC AND CENOZOIC: AN EROSIONAL INTERVAL

Although there is no record of marine rocks younger than Devonian in our area, we suggested that Lower Mississippian strata were deposited here and then removed when a regional uplift in Late Mississippian time initiated a long erosional interval. Interpretation of the geological history of a region when there is no rock record is difficult. We know from the geologic time scale that about 328 million years of earth's history passed during the interval between the Early Mississippian and the oldest Pleistocene deposits. To interpret events during this interval of "lost time," we must use some patches of nonmarine sediments in this area and evidence from other regions where marine and nonmarine deposits are preserved.

We know that there was a general erosional retreat of the Paleozoic rocks down-dip from the crest of the Wisconsin arch during the interval of time prior to Pleistocene glacial deposition. This resulted in an increase in exposure of the Precambrian bedrock and the removal of Mississippian rocks that might have been deposited here. From the rock record elsewhere during the erosional interval in this area, we know there was marine deposition on many of the continents, movement of continental plates relative to the denser underlying rocks, and a continuation of organic evolution. There also were episodes of mountain building that included the formation of the Rocky Mountains, the Alps, and the Himalayas. Great coal swamps covered much of the central and eastern United States in the Pennsylvanian as early forms of plants developed. Amphibians flourished in the Late Paleozoic only to yield their dominant position on land to the reptilian dinosaurs during the Mesozoic. The age of flowering plants, birds, and mammals dawned in the Mesozoic, and by the Cenozoic they dominated the land. Man's ancestors appeared in the Miocene and they evolved into modern man whose desire, but limited ability, to dominate nature provides us with the environmental crisis, the food crisis, and the energy crisis.

In our area there is no record of late Paleozoic amphibians or coal swamps, Mesozoic dinosaurs, Cenozoic mammals, birds, or earliest man. However, we know they existed elsewhere and it is safe to assume that Mesozoic and Cenozoic amphibians, reptiles, mammals, and birds also lived here.

A few thin patches of nonmarine Cretaceous sediments are present in the unglaciated part of southeastern Minnesota and southwestern Wisconsin. These deposits are the iron-rich clay and gravel of the Windrow Formation. Leaf impressions and petrified wood from this formation establish that plants were growing on at least part of this area, while the presence of gravel suggests that streams flowed across this ancient land surface.

Based on these meager bits of information, and from knowledge of events elsewhere, we judge that conditions in the Mesozoic and Cenozoic here really weren't too much different than they are today. However, it is safe to state that the animals present were more attuned to their land. This relatively tranquil scene ended as glacial ice of the Pleistocene began to advance from the north.

PLEISTOCENE: GLACIERS ON THE LAND

The Ice Age

The Pleistocene Epoch of the Quaternary began two million years ago. In our area this was a time of growth, advance, stagnation, and retreat of four major continental ice sheets. At present, about 10% of the world's land areas are ice covered, but during the Pleistocene, glaciers expanded to cover a maximum of 30%. No one can say for sure what caused the earth's climate to gradually become cool and moist enough so that winter snows remained through the summer season to begin a metamorphosis to glacial ice. Only minor temperature changes are required for a buildup of snowfall—perhaps about 5 °C less than our present annual average.

Suggested causes for the swings of temperature required for the ice advances and retreats include differences in output of solar energy, changes in the carbon dioxide content of the atmosphere, dust in the atmosphere, an increase in elevation of some regions, or a change in oceanic circulation. Whatever actually produced the necessary fluctuations in world temperature, snow packs began to remain, and under the weight of annual additions, glacial ice formed. When a critical mass was reached, these great mounds of ice began to flow radially outward from areas of accumulation (Fig. 1-7). They advanced, not in a solid front, but in tongues or lobes up to two miles thick. The outlines of these advancing ice masses are reflected today in the festooned or looping patterns of terminal and recessional moraines. Individual tongues of ice moved at different rates depending on their rate of nourishment and on the resistance of the local preglacial topography.

The advance of great ice sheets on the land was not the only event of note during the Pleistocene Epoch. Accompanying phenomena were major changes in sea level, areas of permanently frozen ground (permafrost) that extended far to the south, a shift in precipitation belts to form large lakes in present desert regions, the blanketing of vast parts of the north-central United States with many tens of feet of wind-borne silts, and a multitude of alterations in the distribution pattern of plant and animal communities.

Sea level depends upon the world's supply of available water. As we mentioned earlier, about 97% of the earth's water is in the oceans, and more than 2% of the remainder is in cold storage as glacial ice. That leaves a meager 1% to fill the lakes and streams, provide the ground water supply, and to moisturize the atmosphere. This is the present water budget of the earth and if we increase any aspect of it we must take away from one or more of the other parts.

With the three-fold expansion of ice caps during the Pleistocene, more water was tied up as glacial ice and the result was a decrease in sea level. Based on evidence of glacial erosion and deposition now beneath the waters of the oceans, we know that sea level was about 330 feet lower than it is today during the maximum period of Wisconsinan glaciation.

Conversely, if the world's ice caps were to melt today, so much water would be released that the level of the oceans would rise about 180 to 200 feet. A shift such as this would create drastic changes in the continental outlines, and inundate all the major port cities of the world. Instead of looking for Atlantis, we could search for New York, New Orleans, Washington and London, to name just a few.

The mass of Pleistocene glacial ice resting upon the earth's crust produced a downwarp of the surface. During maximum glaciation, when most of the continent north of the Ohio and Missouri rivers was covered by an icy mantle at least 10,000 feet thick (Fig. 2-25), the total crustal depression may have exceeded 3,000 feet.

When the ice retreated, the underlying crust slowly began to rebound upward in response to the weight loss. The northern lands of the world are still undergoing this adjust-

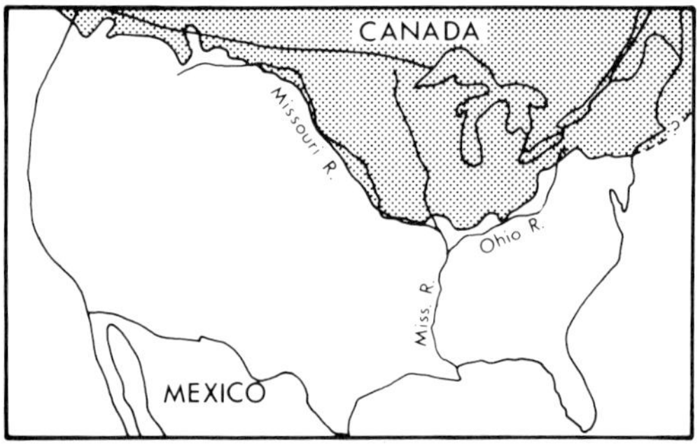

Figure 2-25. Maximum extent of Pleistocene glaciation in the United States. The stippled pattern defines areas which were ice-covered during some part of the Pleistocene. Alpine glaciation in the western United States is not shown. Generalized from various sources.

ment process. At present, the Milwaukee area is rising in elevation at the rate of one-quarter inch a year while Superior, Wisconsin is moving upward twice as fast.

Although the Pleistocene is popularly known as The Ice Age, it lacks the distinction of being the only period of glacial activity in the earth's past. Several earlier ice ages are well-documented and include at least one during the Precambrian and another very widespread episode near the end of the Paleozoic that apparently extended into the early Mesozoic. However, the younger glacial deposits of the Pleistocene are better preserved and more extensively studied.

Our area was overridden by ice lobes from great centers of accumulation in the vicinity of Hudson's Bay. These ice advances extended down into southern Illinois to the approximate confluence of the Missouri, Mississippi, and Ohio rivers (Fig. 2-25). This area of the upper Mississippi River valley contains sediments from more glacial advances than elsewhere and consequently the North American standard classification system for Pleistocene deposits was developed here.

Four major glacial advances took place during the Pleistocene Epoch in North America. Each advance was followed by an interglacial interval, in which climates became similar to today's. The progressively younger Nebraskan, Kansan, Illinoian, and Wisconsinan ice advances are each named for the state where their maximum development is displayed (Fig. 2-26).

Wisconsinan Landscape

Wisconsinan glaciers covered all of upper Michigan, at least three-quarters of Wisconsin, most of northern, and east-central Illinois, much of Minnesota, and they extended into central Iowa (Fig. 2-26). Although the Wisconsinan Stage in Illinois and adjacent areas has deposits dated at 50,000 to 70,000 years before the present, the oldest drift in Wisconsin, according to radiometric and stratigraphic evidence, is generally no older than 30,000 years old. However, recent study of the complex drift in south-central and central Wisconsin, suggests that deposits from an earlier glacial stage may be present in these areas.

At least two major centers of ice accumulation influenced Wisconsinan glaciation. One dispersal area was located west of Hudson's Bay, sending lobes across Minnesota, part of upper Michigan and northwestern Wisconsin from the northwest. Another center was located to the south and east of Hudson's Bay, and these ice tongues advanced by way of the Green Bay and Lake Michigan basins.

The advancing lobes didn't skate along the land surface as crystal blue ice. They lumbered along, continuously incorporating local soil and bedrock within their bulk while leaving portions of their burden upon the landscape as deposits that can all be grouped under the general term *glacial drift.* The transported debris served as a set of abrasive

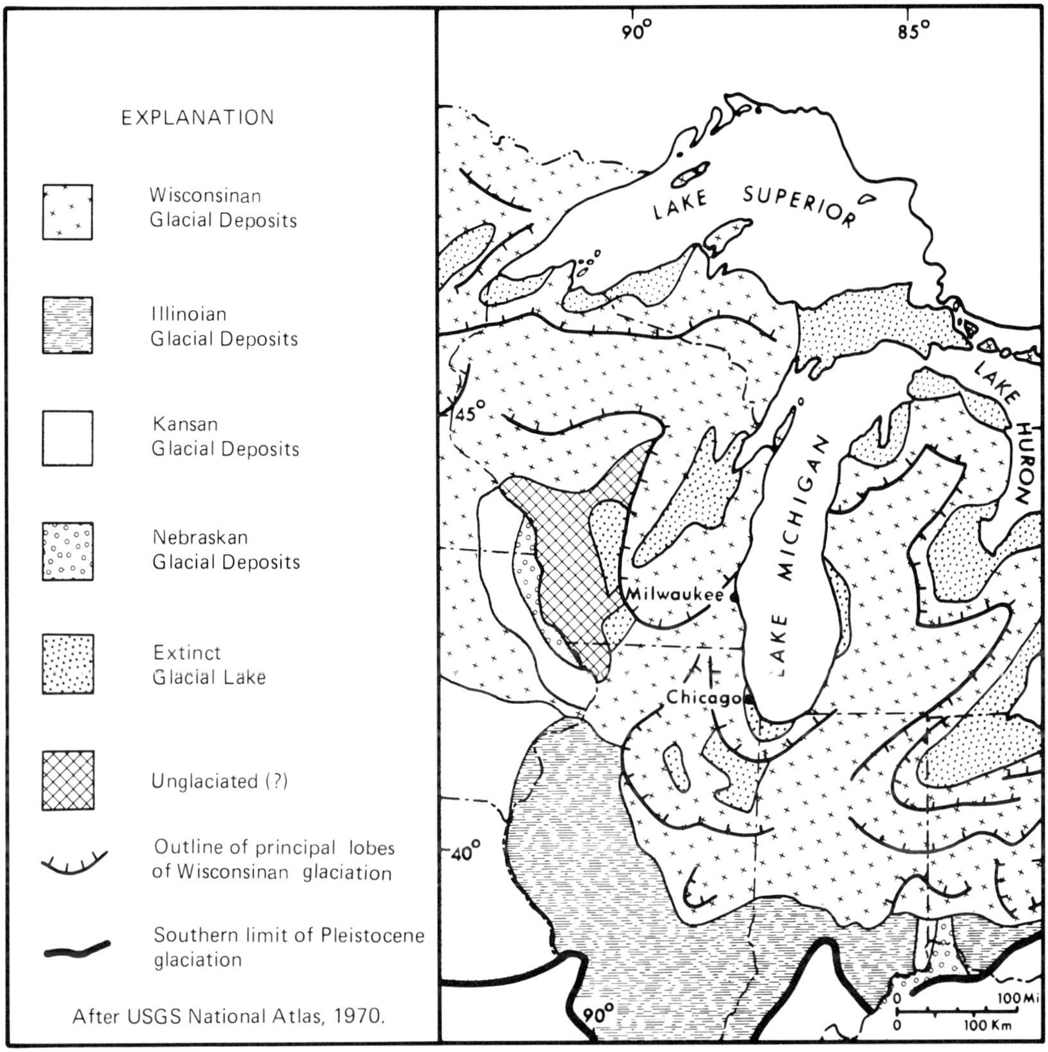

Figure 2-26. Pleistocene glacial deposits of the north-central United States. The extent of the four major glacial advances is indicated. The apparently unglaciated Driftless Area and some of the larger areas of glacial lake deposits are also shown.

tools as the ice moved forward. Advancing lobes ground and polished many bedrock surfaces to a remarkable smoothness, but some tell-tale scratches or *striations* remained to indicate the general direction of passage.

The load carried by the ice was applied to the land in a variety of patterns and situations. If the drift was deposited directly by the ice without the aid of glacial meltwater, the resulting material is called *till* (Fig. 2-27). Till contains particles ranging in size from boulders to flecks of clay. Frequently, the cobbles and boulders in such deposits display striations, produced by abrasive contact with other rocks in the load or with a bedrock surface. The finer particles which resulted from this grinding action are called *rock flour.*

Till deposited in ridges marking the maximum extent of the active portion of the ice lobe, or pauses in its retreat, formed *terminal* (end) or *recessional moraines.* These moraines, mimics of the convex outlines of the glacial lobes, are the major features shown on the glacial map (Fig. 2-28). Where till was indiscriminately deposited as an irregular veneer, it is referred to as *ground moraine.*

An *interlobate moraine* is another type of morainal deposit found here. This variety contains drift deposited by active ice as two neighboring lobes abutted, as well as material derived from stagnation of the debris-laden ice at the edges of the adjacent lobes. Meltwater also funneled into the border area between such ice tongues, and caused an accumulation of water-washed sediments. The impressive Kettle Interlobate Moraine formed between a western lobe moving southward from Green Bay and an eastern lobe moving down what is now the Lake

Figure 2-27. Glacial till is unsorted, ice-deposited material which includes particles that may range in size from boulders to clay. A hammer is shown for scale.

Michigan basin. This complex system of irregular, knobby ridges constitutes the glacial "backbone" of eastern Wisconsin since it extends southwesterly from near Green Bay to Walworth County (Figs. 2-28 and 2-29).

In eastern Wisconsin there are numerous areas with concentrations of distinctive cigar- or whale-shaped hills of till that were formed by the moving glacial ice. These streamlined, elongate mounds are *drumlins* (Fig. 2-30) and they were molded from till, almost as if it were modeling clay, near the outer margins of active or spasmodically re-advancing ice sheets. Rising up to 150 feet above the adjacent drift surface, individual drumlins extend up to two miles in length, and typically are about one-quarter mile wide. Outside of Wisconsin large areas with well-developed drumlin fields are rare, and only northwestern New York and eastern Massachusetts have notable occurrences. In eastern Wisconsin, the drumlin displays in the Lake Mills and Campbellsport areas are especially fine (Fig. 2-28). Since Wisconsin drumlins are so well developed, they should be considered "The Hills That Made Wisconsin Famous."

Drumlins, like bedrock striations, help unravel the history of glacial lobes, since the long axis of a drumlin parallels the direction of ice movement and the blunt, higher end of the drumlin faced the oncoming ice. Today, the iceward ends of many drumlins are tree covered, since farmers often only cultivate the more gently sloping "tails."

The bulldozing action of the ice upon the landscape had the inevitable effect of redistributing bedrock fragments over a wide area. Often, rock varieties obviously "foreign" to a locality were carried there by glacial action from sources hundreds of miles away. The rocks in most till strongly reflect the character of the local bedrock, but distinctive and sometimes geologically rare items were brought southward as *glacial*

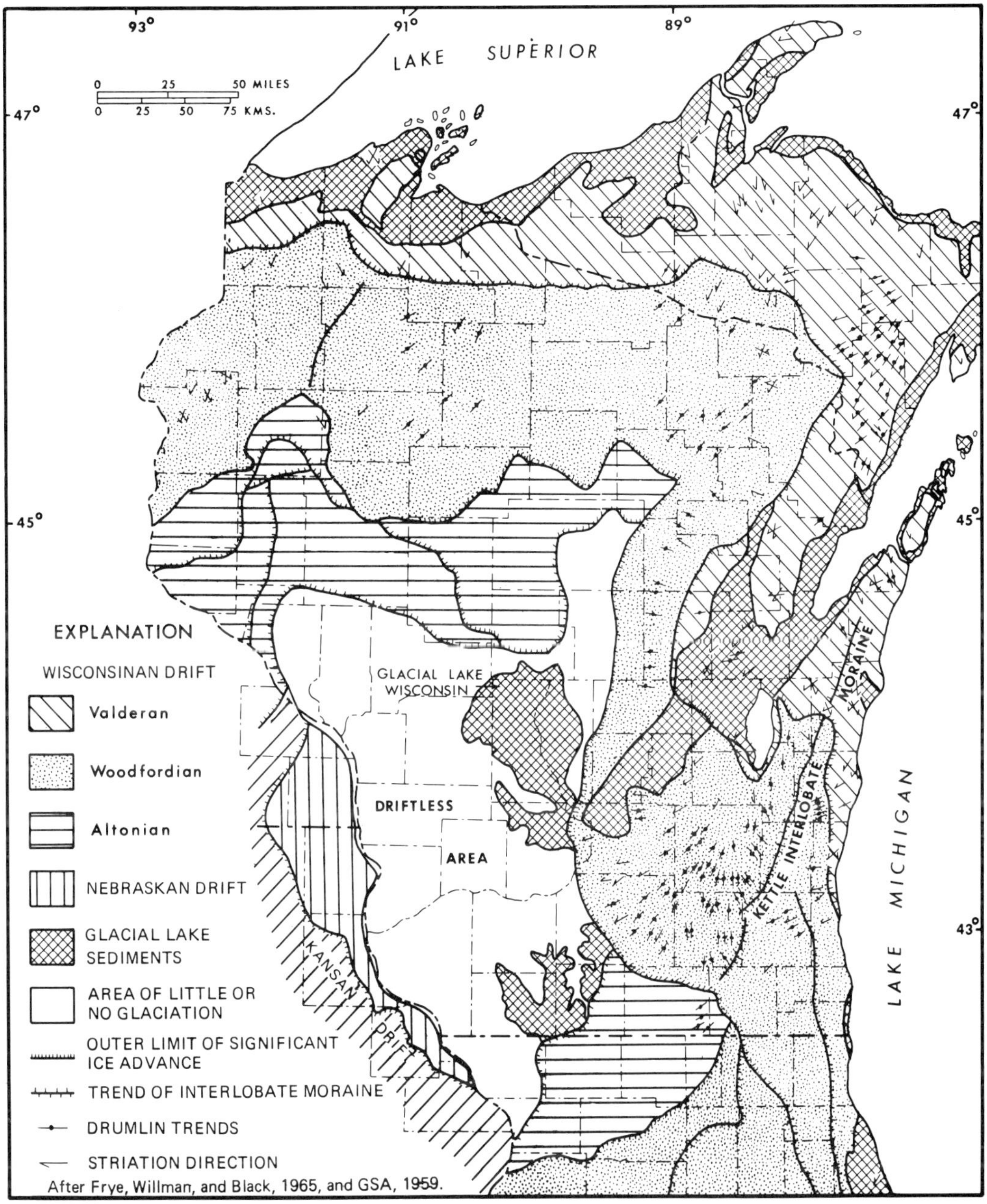

Figure 2-28. Glacial deposits of Wisconsin, northern Michigan, and parts of adjacent states. Major directions of ice movement are determined by drumlin and striation trends. Also illustrated are terminal moraines, the interlobate moraine of eastern Wisconsin, the Driftless Area, and areas with extensive glacial lake deposits.

Figure 2-29. Rolling hills of the Kettle Interlobate Moraine in Sheboygan County, Wisconsin. Wisconsin Department of Natural Resources photo.

erratics (Fig. 2-31). By mapping the locations of a particular type of erratic, the source area can often be determined. This is true of the fragments of native copper which are found in the drift of northern Michigan, Wisconsin, and Illinois. The fan-shaped distribution pattern (or *boulder train*) of copper erratics, some weighing up to two tons, has its apex in the copper country of Michigan's Keweenaw Peninsula.

Diamonds periodically turn up in Wisconsinan glacial drift. As early as 1670, there were reports by pioneer Jesuits of diamonds on some of the Green Bay islands, and approximately twenty were found in Wisconsin during the latter part of the nineteenth century. The largest of these was reported to be over twenty-one carats in weight. Although one can still use the hope of discovery of a drift diamond as an incentive to get the garden spaded, none have been found for more than seventy years. Dear to the hearts of prospectors, geologists, and promoters is the dream of applying the boulder train approach to locate the source of these erratic diamonds. However, as yet no one has done much better than to suggest the diamond-bearing bedrock is "somewhere in central Canada."

Pastoral and solid as her reputation may

Figure 2-30. This long, relatively low, streamlined hill is a drumlin near Campbellsport, Wisconsin. Milwaukee Journal photo.

be, Wisconsin was the site of an early diamond swindle. An enterprising man conceived a get-rich-quick scheme based on the 1846 discovery of a large drift diamond in Waukesha County. After a hard stone of unknown identity was encountered during a well-drilling operation on a farm near Eagle, an unscrupulous Milwaukeean named Colonel Baynton identified the stone as a topaz and purchased it for $1.00. He knew it was a diamond, however, and later sold it in New York to Tiffany and Company for $850.

This was just the beginning for Baynton. He then formed the Diamond Producing Company of Wisconsin, after salting the Eagle area with small diamonds. Stock in his venture sold well until the introduced crystals were discovered to be of South African origin. That ended Colonel Baynton's game, but the 16 1/4 carat Waukesha (or Eagle) diamond went on to reside at the American Museum of Natural History in New York City.

Deposits left by a stagnating, or retreating, ice sheet were not only influenced by the direct action of the ice, but also by the washing and sorting action of floods of meltwater from the diminishing ice. The greater the role that meltwater played, the more highly sorted and stratified the resulting deposits were. Fine silts and clays (rock flour) were washed away in the streams which flowed

Figure 2-31. A rounded glacial erratic of Precambrian granite near Waupaca, Wisconsin.

from the ice margin. This suspended material gave the water the appearance of skim milk. Hence the term glacial milk is applied to such waters flowing from modern glaciers.

Larger rock particles were also affected by the action of running water at and near the ice margin. These *glacial outwash* deposits were rounded and concentrated by size, depending upon the carrying capacity of the

Figure 2-32. This water-washed glacial deposit includes only pebbles and sand in contrast to the till shown in Figure 2-27. A pocket knife in the right center provides scale, and calls attention to a crude, angular alignment of the pebbles. This suggests that the meltwater which deposited the gravel at this place in Walworth County, Wisconsin, flowed from right to left.

water (Fig. 2-32). This activity provided a bonus for the economy of the Upper Midwest. Sand and gravel deposits which were washed and sorted by glacial waste waters form the basis of a thriving industry.

Most sand and gravel deposits in this region originated from streams that flowed outward from the glacier. However, streams flowing on, within, or beneath the stagnating ice left some especially distinctive deposits in areas where the ice didn't advance again. If waters swirled down through cylindrical holes in the debris-laden ice, conical hills of gravel known as *kames* were deposited (Fig. 2-33). If fractures on the ice surface filled with stream-borne gravels were gradually let downward as the ice melted, linear ridges known as *crevasse fills* resulted. Subglacial torrents flowing in ice tunnels deposited layers of sand and gravel to form sinuous, narrow ridges called *eskers* (Fig.

Figure 2-33. A perfectly formed *moulin kame* in Fond du Lac County, Wisconsin. Conical hills such as these are the deposits of debris-laden meltwaters which swirled down through passageways in the wasting ice. The name for this special type of kame comes from early French mountaineers, who thought the swirling motion of meltwater entering such holes in the ice suggested a moulin (mill). Wisconsin Department of Natural Resources photo.

2-34). In some areas, eskers wind for miles as a dry route across a poorly drained post-glacial landscape.

Broad aprons of water-washed debris, which sloped gently away from the ice margin, were the work of meltwater streams carrying a capacity load of sediment. These streams shifted back and forth over earlier deposits to gradually build up a plain of outwash materials. The soils developed on these nearly flat *outwash plains* provide an important part of the tillable farmland of the Upper Midwest.

Figure 2-34. Parnell Esker near Dundee, Wisconsin, in the Northern Kettle Moraine State Forest. This high-standing, narrow, sinuous ridge composed of gravel and sand was deposited by a subglacial stream. The hiking trail along the crest of the esker is one of many trails in this unit of the Ice Age National Scientific Reserve. Milwaukee Journal photo by Stephen C. Liljegren.

As the glaciers continued to recede, the outwash plains dried and wind was free to work on large expanses of sediments until vegetation became established. Wind-deposited glacial silt, or *loess,* blankets the land surface of most of Illinois, Iowa, southern Wisconsin, and a small corner of southeastern Minnesota. Thicknesses are greatest along the Mississippi River where vast areas of glacial outwash were exposed to prevailing westerly winds. The resultant light brown to buff, angular quartz silt deposits are unconsolidated but hold an almost vertical erosion slope.

Sand dunes also formed where large quantities of weathered sandstone or glacial outwash sands were available to the Pleistocene winds. Such deposits accumulated on terraces along the Wisconsin and Mississippi River valleys, along the shorelines of Lake Michigan, and adjacent to some extinct glacial lakes. These dune fields remained active until stabilized by grass and trees.

Where meltwater was impounded by the ice front or against drift ridges, *glacial lakes* or ponds developed. Sediments carried into these quiet water bodies settled out as layers of silt, fine sands, and clays. If conditions allowed calcium carbonate to precipitate, *marl* would form. Many Pleistocene marls contain an abundance of fossil clams and gastropods. Occasionally, large erratic boulders are found in glacial lake deposits. Apparently such rocks were incorporated within icebergs which broke off the ice margin and floated out into the lake before melting.

Most of the Pleistocene impoundments were small and short-lived, but Glacial Lake Wisconsin was a huge body of water that left deposits covering more than 1,800 square miles in central Wisconsin (Fig. 2-28). The maximum water depth may have been as great as 150 feet, based on the elevations of ice-rafted boulders around the margins of this ancient lake.

When blocks of ice broke off a stagnating margin, they often were buried by outwash sediments. After the entombed ice melted, the overlying sediment collapsed and a cavity marked the burial place. These depressions are called *kettles* if they are dry. If they are deep enough to intersect the water table, a *kettle lake* is formed. Many of our scenic kettle lakes are now ringed with water plants, cattails, sedges, willows, and red osier dogwood (Fig. 2-35). This lakeward march of encroaching vegetation forecasts the eventual demise of these small lakes, as the open water gradually yields to aquatic weed-beds, marsh, and then meadow.

The landforms resulting from the wastage of an ice front usually occur in combination. A complex of associated features such as kames, kettles, crevasse fills, and outwash produce a characteristic knob and swale topography.

Glaciation also formed erosional lakes by bulldozing depressions in the bedrock surface. If these depressions extend below the water table or are essentially watertight basins, lakes result. Lakes of this type are common in the northern part of our area where Precambrian rocks are at the surface.

A glance at topographic maps of central

Figure 2-35. Butler Lake, adjacent to Parnell Esker (Figure 2-34), is a small kettle lake. This lake occupies a depression in outwash deposits, which resulted when a buried block of glacial ice melted. Wisconsin Department of Natural Resources photo.

and northern Wisconsin, northern Minnesota, and northern Michigan reveals a high percentage of swampland. Some of these areas were the former sites of glacial lakes but many are not. Glacial deposition caused disturbance, and often complete obliteration, of preglacial stream systems. The postglacial drainage system in this area is less than 10,000 years old and it is still relatively inefficient, so extensive areas of poorly drained land result. These wetlands are now recognized as an important resource, although in the past numerous attempts were made to drain them.

Wisconsin's glacial history and landforms are so uniquely preserved that 33,000 acres have been set aside as the Ice Age National Scientific Reserve. Jointly funded by the state and federal governments, it is under development by the Wisconsin Department of Natural Resources and the National Park Service. The Reserve consists of nine separate units located on or near the edge of the region covered by the Wisconsinan ice advances (Fig. 2-36). Four of the units are also part of the state park and forest systems, but the five remaining units will concentrate on interpretive displays and trails rather than on recreational facilities.

Interstate Park at St. Croix Falls, Wisconsin is a rugged gorge carved in basalt by torrents of glacial meltwater. The St. Croix River now occupies this channel (Fig. 2-37). This park, and a rolling site on the Wisconsinan terminal moraine near Chippewa Falls are the two reserve units within the Northern Highland region of Wisconsin. The characteristics and glacial history of this region are discussed in detail in the next chapter.

The rugged sandstone castles of the Mill Bluff State Park unit near Camp Douglas, Wisconsin, were once islands awash in the meltwaters of Glacial Lake Wisconsin (Fig. 2-38). The Cross Plains unit west of Madison, Wisconsin, is located on the Wisconsinan terminal moraine and includes part of the Driftless Area. Devil's Lake State Park is another unit in the Ice Age National Scientific Reserve. This lake is confined by two dams of glacial debris within a preglacial channel of the ancestral Wisconsin River, which cut through the resistant quartzites of the Baraboo Range.

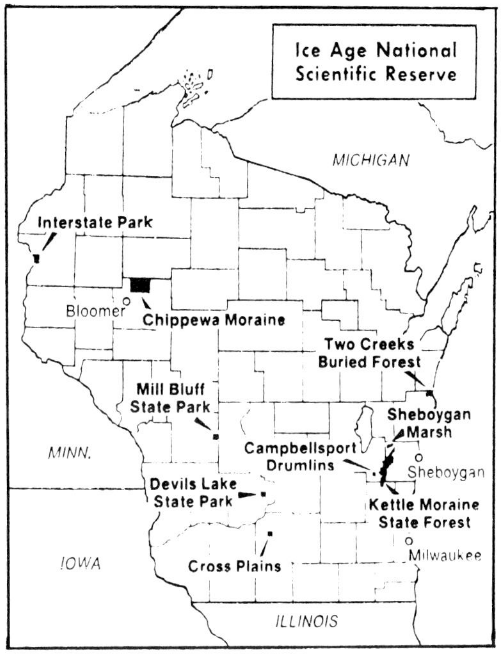

Figure 2-36. Units included in the Ice Age National Scientific Reserve. Map courtesy of The Milwaukee Journal.

Eastern Wisconsin hosts the remaining four units of the Ice Age Reserve. Drumlins are well-displayed at the Campbellsport unit in Fond du Lac County, while the Sheboygan Marsh unit preserves remnants of Glacial Lake Sheboygan. Fallen hemlock and spruce logs sandwiched between two distinct layers of glacial drift at the Two Creeks

Figure 2-37. The scenic Dalles of the St. Croix River at Interstate Park, Wisconsin-Minnesota is a gorge carved through Middle Keweenawan lavas by glacial meltwaters spilling southward to the Mississippi from an ancestral stage of Lake Superior. Wisconsin Department of Natural Resources photo.

Buried Forest site north of Manitowoc, Wisconsin, record the presence of a wooded region which was destroyed 11,850 years ago by the last advance of the ice across Wisconsin. The northern section of the Kettle Moraine State Forest, along the great interlobate moraine of eastern Wisconsin, contains an abundance of magnificent kames, kettles, and other features produced by the wasting ice.

Wisconsinan History

Because of the excellent preservation of the relatively young tills, and the amount of detailed study of the materials and resulting landforms, five substages of the Wisconsinan are recognized. From oldest to youngest these are: the Altonian, Farmdalian, Woodfordian, Two Creekan, and Valderan. The Altonian, Woodfordian, and Valderan represent advances of the Wisconsinan ice sheets while the Farmdalian and Two Creekan record a moderating climate and a recession of the ice (Table VI).

The Altonian Substage began about 70,000 years ago and accounts for about half of the time span assigned to the Wisconsinan Stage. The most recent ice advances of the Altonian moved southeastward across Minnesota and northwestern Wisconsin and westward from the northern Lake Michigan area about 30,000 years ago. These ice

TABLE VI

Summary of Wisconsinan History

Advance	Retreat	Approx. Years Before Present
	Recent	0 to 5,000
Valderan		11,850 to 5,000
	Two Creekan	12,500 to 11,850
Woodfordian		22,000 to 12,500
	Farmdalian	28,000 to 22,000
Altonian		+70,000 to 28,000

Figure 2-38. Mill Bluff State Park, near Camp Douglas, Wisconsin, preserves scenic buttes of Middle Cambrian Wonewoc Sandstone. The flat floor of Glacial Lake Wisconsin is well shown in this photograph. Wisconsin Department of Natural Resources photo.

tongues merged in the center of Wisconsin and may have briefly entered the Driftless Area of southwestern Wisconsin before their withdrawal.

The scene during the Farmdalian Substage was not such an icy one, and a cool, moist climate prevailed from the end of the Altonian until about 22,000 years before the present. Although the glaciers retreated northward, zones of permafrost and buried blocks of stagnant ice managed to survive in many areas of Wisconsin.

The maximum spread of the Wisconsinan ice occurred from 22,000 until 12,500 years ago during the Woodfordian Substage. The terminal moraines marking the maximum area of ice cover during mid-Woodfordian time are distinct enough to be traced from central Illinois, north and westward through Wisconsin, into Minnesota and Iowa (Fig. 2-28). The convex patterns outlining the individual lobes are easily seen on a regional map of glacial deposits (Fig. 2-26). The Kettle Interlobate Moraine between the Green Bay and Lake Michigan lobes continued to develop as the Woodfordian ice wasted. Most of our lakes, including large extinct meltwater impoundments such as Glacial Lakes Wisconsin and Oshkosh, were born during the retreat of the Woodfordian ice. Rock flour and silt, abandoned by overloaded, shifting meltwater streams, were picked up by the wind and deposited as loess over much of the Upper Midwest. Indeed, most of our present landscape dates from this episode in glacial history.

During the gradual warming of the post-Woodfordian climate, soils developed and forest vegetation migrated northward. Retreat of the Woodfordian ice also allowed new drainage outlets for Glacial Lake Chicago, an ancestor of present-day Lake Michigan, to open to the east and the lake level declined. Forests grew along the retreating shoreline on soils which developed on the Woodfordian glacial deposits. This interglacial interval is named for the Two Creeks Forest Bed of eastern Wisconsin, where a buried forest is exposed by erosion along the Lake Michigan shore near the Manitowoc-Kewaunee County line. Logs of spruce, hemlock, and other boreal trees sandwiched between the Woodfordian till and later Valderan lake deposits and tills provide the first record of a forest in Wisconsin since the Altonian Substage. Associated with the tree remains are mosses, sedges, wood-boring beetles, and mollusks.

The Two Creekan forests eventually succumbed to rising lake levels and Valderan ice advances about 11,850 years before the present. Buried logs and stumps at this horizon indicate a maximum tree life-span of 142 years. One stump was found upright in the forest bed with its trunk broken off and aligned in the southwesterly direction of Valderan ice movement.

Additional evidence for multiple glacial advances across eastern Wisconsin is preserved on Silurian bedrock surfaces. Two sets of striations, at about right angles to each other, are exposed on the dolomite bedrock surfaces in the quarry at Valders, Wisconsin (Fig. 2-39). These striations record a southward movement of the Woodfordian ice and a westerly movement of the later Valderan advance.

The advance of the Valderan ice was not as extensive as the Woodfordian event (Fig. 2-28). The Lake Michigan lobe readvanced down the basin carved in soft Devonian shales to the vicinity of Milwaukee. Ice also entered the Green Bay lowland and advanced out of the western end of the Lake Superior basin. These advances limited the Valderan ice to upper Michigan, the far north, northeastern, and northwestern fringes of Wisconsin, and an adjacent area

Figure 2-39. Glacial striations on Silurian dolomite near Valders (Manitowoc County), Wisconsin. The older, deeper striations establish that the Woodfordian ice tongue moved south at this locality while the later Valderan ice advance was toward the west. The hammer handle is aligned north-south.

in Minnesota. In most of these areas the reddish Valderan clay till imparts a distinctive character to the soil.

The source area of earliest Wisconsinan glaciation was probably centered northwest of Hudson's Bay, while later centers shifted eastward toward Labrador. Near the end of the Wisconsinan, a westward shift back toward the earlier source area took place. These shifts in source area, and differing rates of advance and retreat for individual ice lobes makes it difficult to use precise age dates to relate separated ice tongues. As a result, the Pleistocene history of the eastern, northern, and northwestern glacial lobes in this region is generally sequentialized but the precise chronology is not satisfactorily established. For example, the Lake Michigan Valderan lobe is younger than the latest Wisconsinan ice advance into northwestern Wisconsin and adjacent parts of Minnesota.

Large, hardy mammals moved northward with the retreat of the ice front during Two Creekan time, and many remained near the ice margin during the Valderan advance. At this time, our wildlife included bison, mammoth, muskox, caribou, moose, giant beaver, and peccary. The few surviving species, well-adapted to the rigors of a boreal climate, are tundra residents today. Evidence of marine mammals has been found in beach deposits along the shores of the Great Lakes and the St. Lawrence River. Modern day boaters would doubt their eyes if they saw seals, whales, and walrus splashing about.

Humans, lured by the potential for hunting success, were part of the migratory wildlife drawn toward the wasting ice borders. A twentieth century analogy is an expedition planned by big game hunters to the tundra of Alaska and the Yukon Territory. Concurrent with the retreat of Valderan ice sheets about 9,500 years ago is the first evidence of intensive human occupation in the Upper Great Lakes region. This evidence comes from a rock shelter in what is now Sauk County in south-central Wisconsin. Projectile points, well-defined firebeds and artifacts of chert, basalt, bone, teeth, and shell were found here. Apparently the deer hunting mania began early for Wisconsinites since hundreds of split deer bones in the shelter area indicate a fascination for stalking and eating whitetail deer, despite the rigorous climatic conditions which included perennially frozen ground.

Driftless Area: A region of 13,000 square miles in southwestern Wisconsin, a small corner of northeastern Iowa, southeasternmost Minnesota, and part of northwest Illinois appears on most glacial maps as the "Driftless Area" (Fig. 2-28). This region is completely surrounded by glaciated country, although it was not a high-standing land area protruding above a sea of glacial ice. Its character is unlike the surrounding rolling, boulder-strewn countryside. Steep-sided valleys, gently sloping or flat ridges, rocky crags, and an absence of glacial drift create a unique landscape.

The topography, the paucity of glacial deposits, and the presence of residual soil over much of this region indicate a lack of glacial activity and a longer weathering interval than the neighboring areas. However, there has never been total agreement as to the Pleistocene history of this region. Modern researchers cite evidence for at least limited early Wisconsinan ice encroachment upon much of the Driftless Area. Erratic fragments of Precambrian crystalline rocks and Paleozoic sandstone and chert are found on younger bedrock surfaces under loess at numerous localities. To account for these erratics, a partial late Altonian ice cover is suggested by some geologists. Additional evidence supporting Altonian glaciation in this area includes the presence of a thin weathering residuum on the few exposed igneous rocks, absence of any loess older than 30,000 years, and evidence of permafrost conditions.

Why glaciation should have followed a different pattern in southwestern Wisconsin and parts of the adjacent states is not known. As described above, the Driftless Area is not a topographically high region and this eliminates the "island" concept. Subtle topographic differences at the north end of this area may have caused a deflection or a retardation of ice lobes. Perhaps the northern source areas shifted so that advancing tongues missed this area, or maybe glacial advances ended a bit too soon for more than a partial entry. Then again, some combination of time and fortune may have been in control. Regardless of uncertainties about the glacial history of this area, its scenic flavor reminds us what the rest of our region would have looked like without the periodic ice advances of the Pleistocene.

Lake Michigan and Lake Superior: The geologic history of Lake Michigan and the other Great Lakes is a Pleistocene story. This history is complex and differs in detail for each of the lakes since the movements and positions of ice fronts varied throughout the region and this caused outlets to periodically shift or be closed. The underlying bedrock differs and this too exerted an influence on the development of individual lake basins.

Data on the evolution of the modern Great Lakes is mainly derived from the study of abandoned shorelines. Ancient beach and terrace levels, wave-cut cliffs, sand dunes, and lake-bottom sediments all provide clues to this history. The tilting of these shorelines, in response to the mass of glacial ice and to postglacial rebounds, complicates the interpretation of Great Lakes' history. Furthermore, many of these shorelines were reoccupied several times, and this, plus basic disagreements between geologists as to the correct order of occupation, obscures our understanding of details on the origin of the Great Lakes.

The preglacial Great Lake basins were

probably occupied by river systems. As the continental ice fronts began their advances and retreats some two million years ago, these valleys were modified and scoured by glacial deposition and erosion. The entire region was also depressed by the great mass of the ice tongues. This caused the land to tilt northward and meltwaters were impounded between the retreating ice margins and drift deposited by earlier, more extensive, glacial advances. The waters accumulated in these basins until the level reached an overflow route. Ancestors of many of our present day river valleys served as outlets for the developing Great Lakes at various times in the past.

Once an outlet was established, meltwater rapidly cut through the drift which originally held it back until the underlying bedrock was encountered. When this happened, erosion slowed and the outlet was maintained at a relatively stable elevation, until: (a) further retreat of the ice uncovered a lower outlet, (b) the land rebounded sufficiently to cause uplift and abandonment of the outlet, or (c) an ice advance began once again, thus changing the whole scheme of things. Within this complex framework our present Great Lakes progressively changed outlets, from ones that flowed southward into the Gulf of Mexico via the Mississippi River, to an eastward drainage across New York State into the Hudson River, and eventually northeastward into the St. Lawrence drainage system, by both the present-day St. Lawrence exit and a more northerly route that later was abandoned.

The present basin of Lake Superior was developed within the Superior syncline, during a long, chilling Pleistocene history. At the close of Woodfordian glaciation, the first basins to be uncovered were those of ancestral Lakes Michigan and Erie (Fig. 2-40A). Superior probably remained an ice-filled basin until the Two Creekan forests grew in eastern Wisconsin. At this time, an ice-margin (proglacial) lake was formed as the glacial tongue retreated from the southwestern edge of the basin. The outlet for this lake was southwestward down the St. Croix River valley into the Mississippi River (Fig. 2-40B). This ancestral Lake Superior, known as Keweenaw, served as a collecting basin for great thicknesses of red clay and silt that were washed from the iron-rich rock of this region.

The Valderan ice advance blurs the Pleistocene history of the Upper Great Lakes region. It filled all of the Lake Superior basin and much of the Lake Michigan basin as well. At this time the situation was similar to the previous conditions shown in Figure 2-40A. Some of the red sediments of Glacial Lake Keweenaw were rearranged by the advancing ice. As glaciation waned, the western part of the Superior basin collected a group of proglacial lakes which gradually merged into a larger body of water called Glacial Lake Duluth. This lake also drained southwestward down the St. Croix River (Fig. 2-40B).

Lake Duluth persisted through a number of different stages which reflected the downward erosion of the St. Croix outlet. The retreating ice eventually uncovered lower drainage routes southeast across Michigan's Upper Peninsula along the valley of the Whitefish River from near Marquette and Au Train Bay (Fig. 2-40C).

Lake levels slowly subsided as more easterly outlets were utilized and crustal rebound continued. Ultimately the outlet at Sault Ste. Marie, Michigan was used. Erosion of this outlet was rapid until the Superior sill was encountered. This resistant rock became the control of the 602 foot surface elevation of present-day Lake Superior. The Superior basin is the deepest of the Great Lakes with depths that reach 700 feet below sea level.

Like Lake Superior, Lake Michigan has a

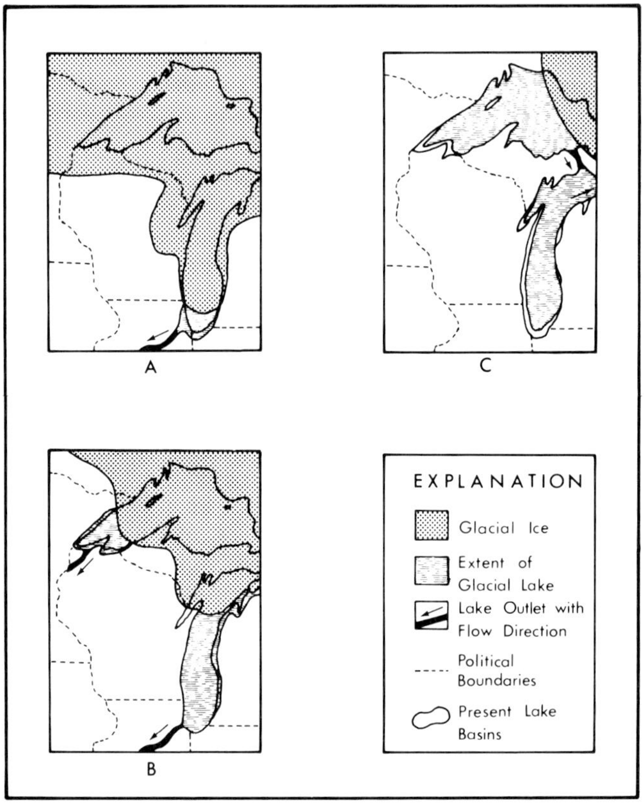

Figure 2-40. Summary of the glacial history of the Lake Michigan and Superior basins. Diagram A shows a retreating Woodfordian ice margin with an early stage of Glacial Lake Chicago developed in the southern part of the present Lake Michigan basin. This lake drained southwest to the Mississippi via the Des Plaines outlet. The Lake Superior basin is still ice-filled. Continued retreat of the Woodfordian ice is shown in diagram B. At this time, a lower stage of Glacial Lake Chicago occupied much of the present Lake Michigan basin and the southwest-draining Des Plaines outlet persisted. A glacial lake (Keweenaw) also existed in the southwestern end of the present Lake Superior basin. This lake drained via the Bois Brule and St. Croix channels to the Mississippi.

As shown in diagram C, a low water stage of the ancestral Great Lakes existed during the Two Creekan. A developing Lake Superior (Keweenaw) established an outlet channel southeast across the Upper Peninsula of Michigan and abandoned the St. Croix outlet. At this time, drainage from the Lake Michigan basin was northeast toward an evolving Lake Huron.

With the advance of the Valderan ice front, a scene similar to that of diagram A resulted again. During this advance, the ice reached the approximate latitude of Milwaukee, the Lake Superior basin was ice-filled and the Des Plaines outlet was utilized once again. Final retreat of the ice in post-Valderan time produced a scene somewhat like diagram C. Glacial Lake Duluth formed in the Superior basin and progressively shifted outlets from the Bois Brule-St. Croix, to the old Upper Peninsula shortcut and ultimately to the area of its present outlet. Ancestral Lake Michigan drained northeast through an outlet in the Straits of Mackinac region.

Postglacial lake stages (Nipissing and Algoma) are not illustrated.

complex history with several chapters missing between the preglacial scene, which featured a large southward-trending river valley, and the time when Woodfordian ice began to recede. This retreat allowed Glacial Lake Chicago to form and expand in the basin being abandoned by the Lake Michigan lobe of Woodfordian ice (Fig. 2-40A). The outlet for this lake was to the southwest by way of the Des Plaines and Illinois rivers to the Mississippi. Lake Chicago existed for a long period of time while several lake stages with different configurations and levels were developed as the ice margin changed. These stages are represented by three "fossil" shorelines. The oldest and highest of these is the Glenwood, which stood about 60 feet above present lake level at an elevation of 640 feet. This stage persisted through a minor late-Woodfordian ice advance.

As the last of the Woodfordian ice retreated, a readjustment of outlets resulted in a lower (Calumet I) stage of Lake Chicago (Fig. 2-40B). The elevation of this stage was about 620 feet. Lake Chicago still drained to the Mississippi, but some of the eastern Great Lakes drained east via the Mohawk-Hudson systems of New York State.

During the Two Creekan interglacial, a low water stage of ancestral Lake Michigan was reached. Crustal rebound temporarily forced the abandonment of the southwestward outlet of Lake Chicago in favor of northeasterly shortcuts toward the St. Lawrence valley. The scene was similar to that shown in Figure 2-40C.

The Valderan advance helped to recreate Lake Chicago's customary outlet to the Mississippi, although the lake also utilized a northeastward outlet until it was closed by ice. The lake again filled to about 620 feet (Calumet II). At this time the lake once again resembled the configuration shown in Figure 2-40A. Eventually, the southwesterly outlet was cut downward 15 feet to a bedrock surface where erosion slowed. This level was the Toleston stage and it was only 25 feet above the present level of Lake Michigan. When the Valderan ice began to retreat, the bedrock-floored outlet preserved the level of the Toleston stage until Lake Chicago merged eastward with early Lake Algonquin in the Huron basin.

As the Valderan ice continued to retreat northward, crustal rebound and newly opened eastern outlets resulted in another low water stage in many of the lakes (Fig. 2-40C). This predecessor of Lake Michigan was the rather insignificant Lake Chippewa.

The Nipissing stage of the Great Lakes developed during postglacial time as crustal rebound slowly elevated the outlets and allowed a gradual rise in water level to that of the earlier Toleston beaches. At this level, ancestral Lake Michigan overflowed its southwestern outlet for the last time. Winds along the Nipissing shoreline produced large areas of sand dunes, which were best developed on the northern, southern, and eastern sides of the lake.

Postglacial rebound of the earth's crust, stabilization of amounts of available water, the elimination of some outlets, and downcutting of others resulted in the Algoma stage at an elevation of 595 feet. Gradual changes resulted in modern Lake Michigan with an elevation of 580 feet about 2,500 years ago.

Bathymetric charts of Lake Michigan reveal two deep basins separated by a divide area of resistant Middle Devonian bedrock and possibly some terminal moraine material. The northern basin is quite irregular, suggesting localized glacial scour of different types of beckrock. The deeper (923 feet), more regular, southern basin is cut in relatively soft Upper Devonian to Lower Mississippian shales. At its deepest point, Lake Michigan is 343 feet below sea level.

Glacial Heritage: Our glacial heritage has

left the Upper Midwest with many benefits more tangible than those of an intellectually stimulating history. The continental ice lobes, in their bulldozer-like advance across the north-central United States, had a measurable effect on agricultural resources. They scraped off high areas, filled in the lows, and blanketed vast regions with water-washed or wind-blown glacial sediments. As a result, much of this area is suitable for large-scale machine agriculture. Despite the fact that farmers for more than a hundred years have cursed the erratic boulders in their fields, which must be carted off to fence rows or central piles, the general quality of soils has been enriched by the new minerals that were transported and mixed by glacial action. Wind-blown glacial silts were blanketed over relatively infertile bedrock surfaces in the Driftless Area, improving agricultural conditions even in this apparently unglaciated region.

The grinding of ice over bedrock and the later sluicing action of meltwaters on glacial debris produced widespread deposits of sand and gravel. These accumulations lack the mystique and adventure-packed history of placer gold deposits but they are nonetheless an extremely valuable economic resource.

Glacial sand and gravel deposits also serve as valuable water reservoirs for many small communities and rural property owners. They often provide good quality water at shallow depths, thus eliminating the need for expensive bedrock wells.

The landforms created by glacial erosion and deposition have contributed immeasurably to the scenic and recreation potential of this area. Thousands of beautiful lakes in Wisconsin, Minnesota, and northern Michigan, including Lake Michigan and Lake Superior, owe their existence to the Pleistocene. Many of the region's most popular recreation areas, state parks, and forests are the legacy of the icy past.

The infinite variety of plant and animal communities in this region also reflects our glacial history. Plants of the boreal forest still linger almost as far south as the Wisconsin-Illinois state line while many of the animals that lived along the edge of the glacial ice remain as residents.

A contemporary Pleistocene specialist, Professor Robert F. Black, who helped to unravel Wisconsin's geologic history, has written that the events of the past 30,000 years are more important to Wisconsin's scenery and economy than all the previous billions of years combined. After some reflection, we feel that he was not merely boosting his academic specialty but speaking with great wisdom.

Chapter 3

NORTHERN HIGHLAND

CONCEPT OF PHYSICAL PROVINCES

The terms Canadian shield, Superior Upland, and Central Lowland were used in Chapter 2 to set our area on the geological stage (Fig. 2-2). These labels result from a geologist's desire to divide the physical landscape into regions with related characteristics for more effective study and communication. Various criteria can be used in this pigeonholing process, but our grouping emphasizes aspects of geology and topography.

The geology of an area, including the type and structure of the bedrock, is a basic factor reflected in the landscape. A massive granitic rock will not erode as easily as a poorly cemented sandstone, and landforms that develop on flat-lying sedimentary rock will be different from those in structurally complex areas. Since topographic controls such as these are inherited from past events, they reflect the geologic history of an area.

The climate of a region is also important in defining a physical province, since it determines the type and intensity of the weathering and erosional processes. Given the same geologic setting, different climatic conditions operating over a sufficient period of time will produce dissimilar landscapes. A dry climate tends to accentuate bedrock structure. The lack of moisture inhibits chemical weathering and the growth of vegetation, thus retarding soil formation. The lack of vegetation also allows infrequent rainfall to remove weathered material very efficiently. A warm and moist climate, on the other hand, disguises the underlying bedrock with the rapid development of soil and vegetation. In the first situation the land appears rugged and angular, while in the second example, the topography is much more subdued.

Even the influence of prehistoric weather conditions is important. The climate which resulted in the buildup and advance of the Pleistocene continental ice sheets dramatically changed the landscape of the glaciated areas. The masses of spreading ice transported vast amounts of underlying soil and bedrock to adjacent regions before depositing them to create a new landscape. Because most of the north-central United States was covered by continental glaciers only "yesterday," much of the bedrock geology is obscured by glacial deposition. The dramatic contrast between glaciated and nonglaciated landscapes is well-illustrated along the margin of the Driftless Area in southwestern Wisconsin and parts of adjacent states (Fig. 2-28).

As you might guess, the dividing and grouping process involved in defining physical provinces is not a cut-and-dried scientific exercise. The basic problem with any such classification is that mere mortals are imposing an arbitrary organizational scheme upon the natural world. Consequently, the results will not always be perfect. Sometimes the boundaries between natural areas are sharp and distinct, but in many instances they are not. For example, the eastern edge of California's rugged Sierra Nevada Mountains (a part of the Sierra- Cascade Province) forms a spectacular province margin that rises 6,000 to 8,000 feet above the adjacent Great Basin to the east (a part of the Basin and Range Province). However, the boundary between the Basin and Range Province and the Rocky Mountains is not clearly defined, since it occurs within a complex area of mountains, basins, and plateaus.

In spite of such difficulties, the establishment of natural or physical provinces with similar geologic frameworks and topographic expressions is the best way to divide a region into areas which can conveniently be analyzed and described. In our region, we recognize four physical provinces, and in one of these, we consider several small lowland areas as subprovinces (Fig. 3-1).

The bedrock in our largest province, the *Northern Highland,* is Precambrian, although some of the lowland areas shown in Figure 3-1 are underlain by sandstone that may be Lower to Middle Cambrian. This province is part of the Superior Upland which is a southward extension of the Canadian shield (Figs. 2-2 and 2-4). The other three provinces are parts of the Central Lowland and they are underlain by nearly horizontally-bedded Paleozoic sedimentary rocks that flank the Northern Highland (Figs. 2-4 and 3-1). The *Central Plain* is a lowland developed on Cambrian sandstone on the southern margin of the Northern Highland (Figs. 2-4 and 3-1). The *Western Uplands,* of western and southwestern Wisconsin, and the *Eastern Ridges and Lowlands,* paralleling the Lake Michigan shore, are the remaining two physical provinces. The bedrock in these provinces consists of resistant and less resistant sedimentary rock units which form uplands or ridges that alternate with valleys or lowlands. We extended these two province names into northern Michigan, northern Illinois, and adjacent parts of Iowa and Minnesota (Figs. 2-4 and 3-1). In some of these states, these provinces are divided by others into smaller physical areas.

GENERAL CHARACTER

The oldest rocks in our area are exposed in the Northern Highland Province. This gentle upland of igneous and metamorphic rocks, with a cover of glacial deposits, is the southernmost extension of the vast Canadian shield (Figs. 2-1 and 2-3). Although the bedrock is primarily of Precambrian age, Paleozoic seas deposited a thin veneer of sedimentary rocks over this ancient surface. Postdepositional upwarping of this region produced the Wisconsin dome and caused the Paleozoic rocks to slope away from the center of uplift in Vilas and Oneida counties (Figs. 2-4 and 2-16). After uplift, the Paleozoic rocks eroded downdip to the south, east, and west to produce a concentric system of lowlands separated by dipping ridges (cuestas) surrounding the Northern Highland. This lowland and cuesta pattern reflects variations in the erosional resistance of different sedimentary rock units. The province boundary of the Northern Highland closely follows the contact between the Precambrian crystalline rocks and the retreating, erosional edge of Upper Cambrian sandstone.

True to its highland label, the land slopes away from an elevation of about 1,700 feet above sea level. The highest points in this region are generally resistant bedrock hills on the upland surface. Mt. Curwood in Baraga County is a hill of this type, and with an elevation of 1,980 feet, it is the highest point in Michigan. However, Tim's Hill in Price County is capped with glacial deposits and it soars to the Wisconsin state record of 1952.9 feet!

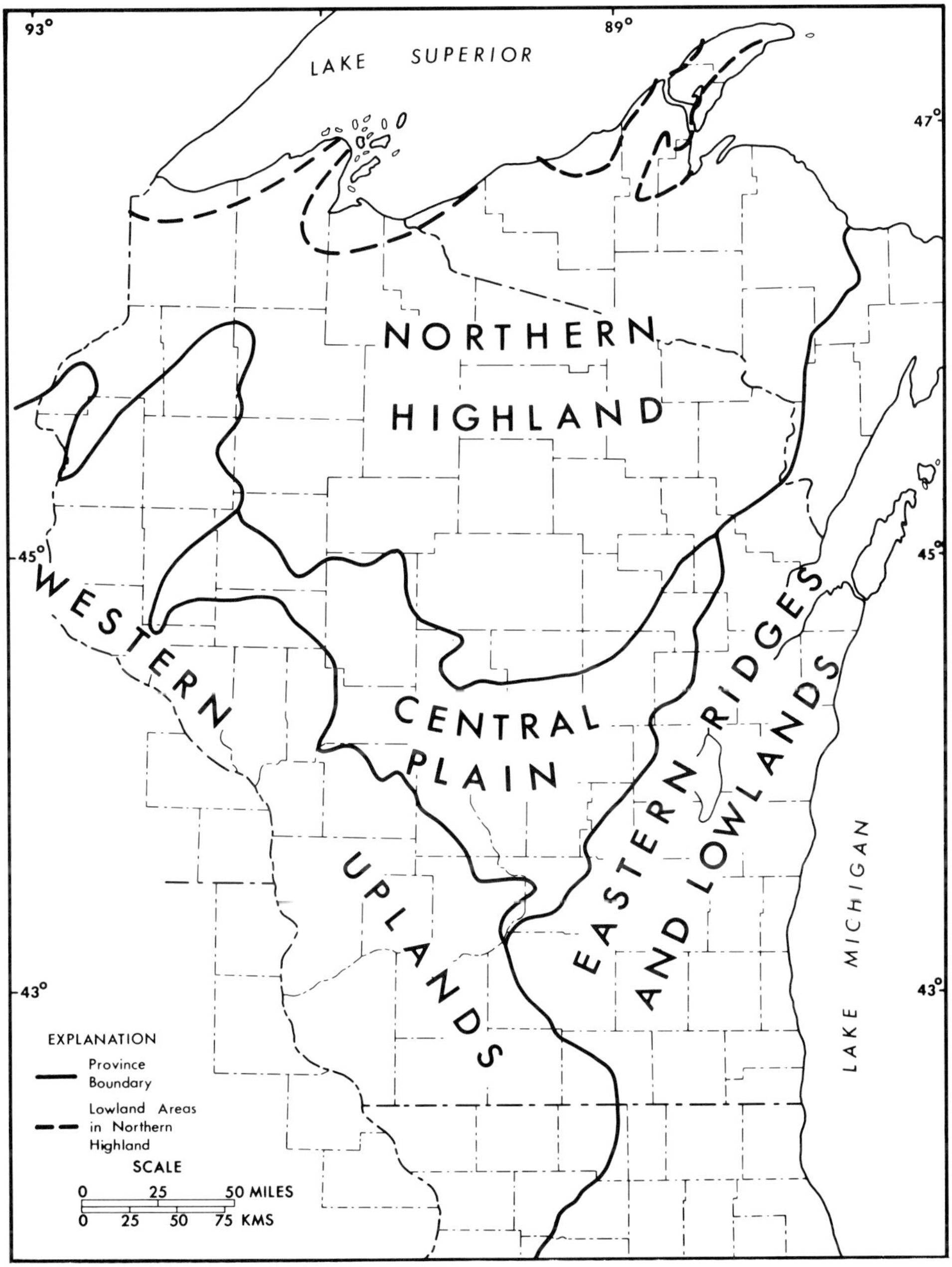

Figure 3-1. Physical provinces of Wisconsin, northern Michigan, and parts of adjacent states. The lowland areas bordering Lake Superior within the Northern Highland Province are outlined by dashed lines. Modified from Martin, 1932 and Black, 1964.

All our major river systems originate in this upland region, and flow into Lake Superior, Lake Michigan, or the Mississippi River. Two parts of the Northern Highland are densely dotted with glacial lakes. One of these is in northwestern Wisconsin and it extends northeasterly from Polk County to Bayfield County. The other is centered in Wisconsin's Vilas and Oneida counties. Few parts of the world contain so many lakes with so little intervening land. Oneida County alone has more than 800 of Wisconsin's 9,000 lakes, and hundreds of bogs and marshes mark the sites of former lakes.

The northern parts of Douglas, Bayfield, and Ashland counties, Wisconsin, have a somewhat different physical and climatic character than the rest of the Northern Highland Province. Consequently, this area along the Wisconsin shore of Lake Superior is sometimes treated as a separate province called the Lake Superior Lowland, although we consider it a subprovince within the Northern Highland. The Lake Superior Lowland is interrupted by the relatively high-standing, rocky Bayfield Peninsula, but it continues as a low plain developed on Late Precambrian sedimentary rocks around Chequamegon Bay at Ashland, Wisconsin (Fig. 3-1). From there, it continues northeast along Lake Superior into northern Michigan as a narrow band that is briefly interrupted by the Porcupine Mountains. A somewhat similar lowland, underlain by Lower to Middle Cambrian (?) Jacobsville Sandstone, forms the southeast half of the Keweenaw Peninsula and it extends inland from the area occupied by Keweenaw Bay (Figs. 2-4 and 3-1).

The lowlands rise several hundred feet above present lake level as gently sloping plains that were covered by glacial lakes ancestral to Lake Superior (Fig. 2-40). These lakes left a legacy of red clays that obscure the bedrock and limit agricultural productivity. However, this liability is somewhat offset by the lake-margin climate which is moderate, and allows a longer growing season than the rest of the Northern Highland.

The margins of two of the lowlands are partially defined by major faults. The Keweenaw fault forms the northwestern boundary of the lowland along the Keweenaw Peninsula of Michigan (Fig. 3-2). In Wisconsin, the Douglas fault forms part of the southern border of the Lake Superior Lowland in Douglas County (Fig. 2-3). This fault, which brings Keweenawan basalts up over Lower to Middle Cambrian (?) sandstones, forms an escarpment that rises several hundred feet above the lowland. Streams flowing northward off this escarpment cascade down the resistant rocks along the fault line to form scenic waterfalls such as Manitou and Amnicon Falls. Below the falls, gorges are carved in glacial lake clay and sandstone bedrock of the lowland as the streams continue on to Lake Superior.

The Lake Superior shoreline is more dramatic in appearance at the end of the Bayfield Peninsula in Wisconsin and along the northwest side of the Keweenaw Peninsula of Michigan. On the Bayfield Peninsula, the shoreline is cut in reddish-brown latest Precambrian to Cambrian (?) sandstone rather than in glacial lake or drift deposits. Sea cliffs, arches, and pillar rocks provide an excellent foil for pine and birch which cling precariously to them. The Bayfield Peninsula points offshore to the Apostles, a rocky flotilla of 22 islands (Fig. 3-2). These islands, excluding Madeline and Long islands, form the nucleus of the Apostle Islands National Lakeshore that eventually will incorporate about 42,000 acres of island and mainland shore.

Late Precambrian conglomerate and interbedded basaltic lava outcrop along the northwestern shore of the Keweenaw Peninsula. Variations in the erosional resistance of these steeply dipping rock layers create a scenic shoreline, as rocky headlands alternate with small bays.

The Northern Highland is portrayed by local Chambers of Commerce as the land where the legendary muskie lurks in a conifer-ringed lake of the cool north country. This vision rather effectively illustrates that the major economic resource of this area at present is tourism. Most of the iron and copper mines which once flourished are played-out and closed. Rusting headframes,

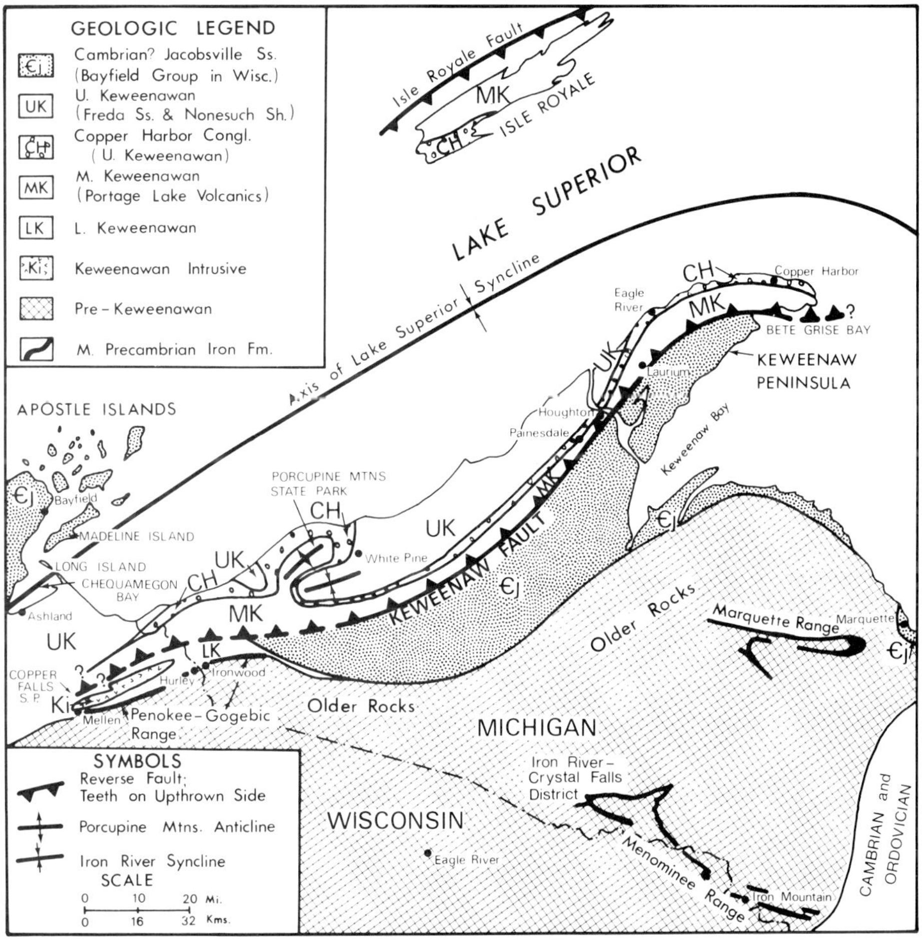

Figure 3-2. Generalized geologic map of part of northern Michigan, Isle Royale, and adjacent parts of Wisconsin. Featured are the Lake Superior syncline, Keweenaw Peninsula, Keweenaw fault, Isle Royale fault, Bayfield Peninsula, Apostle Islands, Chequamegon Bay, Porcupine Mountain anticline, and the three iron ranges of Michigan and Wisconsin.

overgrown mine dumps, and towns past their prime recall more prosperous times for the mining industry (Fig. 3-3). Although 6 iron mines are still active in northern Michigan, only one mine now produces copper. However, better times for the miners may be on the way. A relatively new iron mine is producing from Precambrian rock immediately south of the Northern Highland border at Black River Falls, Wisconsin. Increased exploration activity is responsible for the location of 3 new copper-zinc ore bodies in northern Wisconsin, and several copper mines in the Keweenaw Peninsula may reopen.

Most of the Northern Highland is forested, despite the complete cutover of white and red pine in the 1890s, Paul Bunyan, and several major forest fires. Reforestation programs since the end of World War I have helped to restore much of the timber, and now large areas are set aside as state or national forest. Forest reserves at present are adequate to support a healthy pulp and lumber industry.

Agriculture in the Northern Highland is limited by a short growing season, generally poor soils, irregular topography, and the lake-swamp environs of this area. However, dairying is still an important occupation.

Figure 3-3. A rusting headframe rises above an abandoned underground mine near Pence, Wisconsin on the Penokee-Gogebic Range.

BEDROCK AND STRUCTURE

Our oldest and most complexly deformed rocks underlie the Northern Highland Province. Scattered exposures, mine shafts, drill holes, and geophysical data (gravity and magnetics) provide some information on the bedrock. As shown in Figure 2-9, pre-Keweenawan Precambrian rocks are older at the northern edge of this province than they are to the south. However, recent information indicates that a gneissic complex at Stevens Point, Wisconsin may be as old as 3 billion years. Obviously, we still have things to learn about the Early and Middle Precambrian history of the Northern Highland.

As previously described, the event-filled geologic history of this province includes repeated episodes of mountain building, volcanism, inundation by seas, sedimentation, and erosion. Once the site of lofty mountain ranges, a long erosional interval beveled the region to a low plain with moderate relief. Only scattered ridges and isolated hills of more resistant quartzite and basalt escaped the relentless powers of erosion and remained as topographic highs. Diverse rock types that include granite, gneiss, basalt, quartzite, marble, slate, and schist are now largely covered by glacial debris, dense vegetation, and swamp.

Low ridges of Middle Keweenawan basalt are exposed in northwestern Wisconsin, and they extend across northern Michigan onto the Keweenaw Peninsula, where they are copper-bearing (Fig. 3-2). Basaltic ridges of similar age are present along the Minnesota shore of Lake Superior and on Isle Royale. In all instances, they are the result of great outpourings of lava during the Late Precambrian. A thick sequence of Late Keweenawan to earliest Paleozoic (?) conglomerate and sandstone overlies the lava series. The aggregate thickness of Keweenawan volcanic and sedimentary rocks may reach 50,000 feet in the Lake Superior syncline.

Elsewhere in the Northern Highland, a few ridges and hills protrude above the old erosional surface. Typically composed of Animikean quartzite, these erosional remnants are called *monadnocks.* Flambeau Ridge (Chippewa County), Powers Bluff (Wood County), Thunder and McCaslin mountains (Forest, Marinette, and Oconto counties), Rib Mountain (Marathon County), the Barron Hills (Barron, Rusk, and Sawyer counties), and the Marquette, Menominee, and Penokee-Gogebic iron ranges rise an average of 300 feet above the surrounding countryside. All of these features are described in more detail below.

The Precambrian rocks of the Northern Highland formed the basement upon which the younger sedimentary rocks were arranged. After a Late Precambrian to Early Paleozoic erosional interval affected most of our region, the Late Cambrian seas transgressed, depositing conglomerate and sandstone over the eroded basement complex. Younger Paleozoic seas also covered this province and additional sediments were deposited. Later in the Paleozoic, the Northern Highland was gently upwarped and the sedimentary rocks were removed by water and wind. Isolated *outliers* of sedimentary rock along the border zone of the Northern Highland testify to the existence of a more extensive Paleozoic cover.

Beyond the province borders of the Northern Highland, some patches of Precambrian rock are exposed where stream erosion has cut through the overlying Paleozoic rocks. An example is the Precambrian rock that lies along the 40-mile segment of the Black River of Wisconsin, between Hemlock (Clark County) and Black River Falls (Jackson County) (Fig. 2-4). Such exposures are called *inliers.* Elsewhere in our region, drilling through the Paleozoic cover establishes that the Precambrian basement is just as lithologically and structurally complex as it is on the outcrop.

GLACIATION

The glacial history of the Northern Highland is complex, since this area was traversed by multiple lobes from at least two major centers of ice accumulation. Glacial fingers advanced from the northwest, north, and northeast out of what are now the Lake Superior and Lake Michigan basins. After millions of years of pre-Pleistocene erosion, the entire region was greatly modified by glacial erosion and deposition (Fig. 2-28).

The central part of the Lake Superior syncline was especially vulnerable to glacial erosion because the Late Keweenawan to Middle Cambrian (?) sedimentary rocks provided little resistance to the lumbering masses of ice. In addition, ridges of Middle Keweenawan basaltic lava on the peripheries of the syncline helped to confine the ice. The result was that hundreds of feet of rock were removed by glacial scour, and the Lake Superior basin was cut 700 feet below sea level (Fig. 2-10). After the glaciers retreated, this basin was occupied by Lake Superior, the deepest of the Great Lakes.

Early Wisconsinan (Altonian) ice covered most of the Northern Highland, but later advances across this province extended progressively less far south. As a result, there are four areas of different glacial character within the Northern Highland. One of these is a small rectangular region that projects north along the Wisconsin River valley to beyond Wausau. Although this country is assigned to the Driftless Area in Fig. 2-28, recent work indicates it was glaciated by a pre-Altonian advance. It also received quantities of outwash during later ice advances. In higher areas, residual soils developed on the bedrock, and a dendritic drainage pattern became so well-established that lakes and swamps are uncommon. Rib Mountain and Mosinee Hill, however, survived as quartzite monadnocks that rise above a gently rolling landscape.

The Altonian glacier advanced as far south as Clark, Wood, and northern Marathon counties in central Wisconsin. Here, the drift varies from 5 to 170 feet in thickness, but more than 30,000 years of post-Altonian weathering have muted the glacial landscape. Better drained and less rugged than regions of younger drift to the north, this area has the best farmland in the Northern Highland.

Many other parts of the Northern Highland are thickly blanketed by younger glacial drift deposited during the maximum stages of the Woodfordian advance (Fig. 2-28). Prominent terminal moraines associated with this glaciation are easy to trace across the northern landscape. Knobby and kettled in many stretches, the glacial till is composed primarily of debris from the igneous and metamorphic rocks of the Northern Highland.

The topography and drainage of areas underlain by Woodfordian deposits are in marked contrast to those with older drift. The rugged, poorly drained landscape reflects the shorter period of time available for erosion and for the establishment of an efficient drainage system. Swamps and lakes abound, particularly in the vast regions of pitted outwash. The youthful drainage network is incompletely developed, and rapids and small falls are common.

Two interlobate moraines were created between adjacent Woodfordian ice lobes within the Northern Highland. One formed between ice moving out of the Lake Superior basin and the Lake Michigan ice. This morainal system trends southwest from Dickinson County, Michigan to Langlade County, Wisconsin. The Antigo outwash plain in Langlade County, Wisconsin is just south of this interlobate moraine and it is one area affected by Woodfordian glaciation that provides good agricultural land. Another interlobate moraine formed in northwestern Wisconsin at the junction between two lobes of ice moving out of the Superior basin. This moraine trends southwest from southern Bayfield County to northern Washburn County (Fig. 2-28).

The more recent Valderan ice advance modified most of northern Michigan, parts of northeastern and northwestern Wisconsin, and a narrow band that extends south along Lake Michigan to Milwaukee. Many Valderan deposits are thin, irregular veneers on the thicker Woodfordian drift. Two Creekan buried forests are found between the Woodfordian and Valderan in eastern Wisconsin and Iron County, Michigan, but no pronounced weathering profile separates these tills in the Northern Highland.

Although the Valderan ice was thinner and weaker than its predecessors, it formed a well-defined terminal moraine about 5 miles wide and with about 100 feet of local relief in Douglas County, Wisconsin (Fig. 2-28). This is the Nickerson Moraine, and it sprawls westward for some 20 miles into Minnesota before it merges with another morainal system. Other well-defined parts of the Valderan terminal moraine are present in Iron County, Michigan.

In addition to glacial deposition, much of the Northern Highland was subjected to considerable erosion. Most of the bedrock projecting through glacial till is rounded, polished, and striated. In many areas, the bedrock was scoured deeply enough to provide lake basins.

DRAINAGE

The Northern Highland is the headwaters region for the major river systems of our area. From this area of abundant water, stored in lakes and swamps, these rivers flow southwest, south, southeast, and north from a center near Land O' Lakes, Wisconsin. Tributaries of the St. Croix, Flambeau-Chippewa and the Wisconsin flow to the Mississippi River. Green Bay or Lake Michigan are the destinations of the Escanaba, Menominee, Peshtigo, Oconto, and the Fox-Wolf rivers. A series of short streams, including the Nemadji, Bois Brule, Bad, Montreal, Presque Isle, and Ontonagon follow uncomplicated patterns north toward Lake Superior (Fig. 3-4).

The Wisconsin River flows south from headwaters near Lac Vieux Desert in northern Vilas County, Wisconsin. It drains 22% of Wisconsin before entering the Mississippi

Figure 3-4. The Bois Brule River, Bayfield County, Wisconsin. From the mid-1600s well into the 1800s, this river was part of the favored canoe route of French and English explorers, trappers, missionaries, and traders between Lake Superior and the Mississippi River. The Bois Brule rises near the headwaters of the St. Croix River, a major tributary of the Mississippi, at Solon Springs, Wisconsin. A short portage connects these two waterways. This route follows the ancient spillway of Glacial Lake Duluth. The flood of meltwater that passed here enlarged the valley now occupied by this beautiful stream. Wisconsin Department of Natural Resources photo.

River at Prairie du Chien. North of Merrill, the Wisconsin flows through an irregular landscape blanketed by young glacial drift, and a number of rapids interrupt the river's southward course. The largest of these is at the site of Grandfather Dam and power station, 10 miles north of Merrill, where the river drops 90 feet in a mile and a half. These rapids provided challenges in the days of the early log drives. In fact, the largest log jam in Wisconsin history occurred at Grandfather Falls, when 80 million feet of timber became entangled as a mass of logs 20 feet in height. South of Merrill, glacial drift is either older or absent, and the Wisconsin flows in a steep-sided channel cut into outwash gravels. In this area, tributary drainage is efficient, and there are very few lakes.

Despite problems such as log jams, the lumber industry made extensive use of almost all the Northern Highland rivers. Logs were harvested in winter and rafted down the rivers during spring runoff to sawmill towns that were often situated at impassable rapids. Many of these rapids are now dammed to provide electrical generating sites. There are so many impoundments with hydroelectric plants along the Wisconsin River that it is considered one of the nation's "hardest working" rivers.

The two greatest concentrations of lakes in the Northern Highland are in areas where glacial deposits are young and the drainage system is poorly developed. One is centered in Vilas and Oneida counties of Wisconsin, and Gogebic County, Michigan. This is the highland lake district and it is at the headwaters of the Wisconsin, Flambeau, Ontonagon, Wolf, and Menominee rivers. The northwestern lake district of Wisconsin (Washburn and Burnett counties) is at the headwaters of the St. Croix and Chippewa rivers.

SPECIAL FEATURES

The Northern Highland abounds in special geologic features that are worth visiting. We have selected the best and most accessible for discussion.

Some of the features described, like lakes, waterfalls, and quartzite monadnocks, are widely scattered through the region. Others are state or national recreation areas. Chequamegon Bay-Apostle Islands (Wisconsin) is a national lakeshore on Lake Superior. Interstate Park (Wisconsin-Minnesota) and Porcupine Mountains (Michigan) are state parks, and Isle Royale is a national park. Three historic iron ranges (Penokee-Gogebic, Marquette, and Menominee) in northern Wisconsin and Michigan are also highlighted. Another selected locality is the entire Keweenaw Peninsula of Michigan. Many of these features are visited on a Lake Superior field trip that is described in a companion volume.

Lake Districts

Most visitors to the Northern Highland come because of the scenic lakes and the fish within them. The northwestern and highland districts, described above, contain major concentrations of lakes. Among the largest and best known of the northwestern lakes are the Namekagon, the St. Croix and Chippewa flowages, and Lac Court Oreilles.

Lake density in the highland district is such that 40% of the area is either water or swamp. This is a record that is only equaled by similar glaciated areas in southern Finland, and parts of Ontario and northern Minnesota. Among the larger and better known lakes in this district are Lac du Flambeau, the Flambeau Flowage, and Trout, Tomahawk and Big Arbor Vitae lakes (Fig. 3-5). The Flambeau and

Manitowish flowages cover areas of 15 to 25 square miles and include a tangle of lakes, wooded waterways, and swamps.

Our northern lakes vary greatly in size, depth, and regularity of outline. Some are connected by wandering channels while others exist in solitary splendor. Regardless of such individual variations, these lakes are glacial in origin. Many are kettles, others fill irregular depressions in the ground moraine, some were created by the damming of previous drainage systems by morainal debris, and still others formed in depressions scoured out of the bedrock. Swamps and marshes are commonly associated with such lakes. Some occupy depressions of glacial origin that were not deep enough to form lakes, but many others were lakes which are now filled with mats of vegetation and sediment. All the lakes and swamps reflect deranged drainage in a recently glaciated region. Utilizing an ordinary highway map, it is possible to delineate the boundary between older and younger drift in the Northern Highland by tracing the margin of the lake concentrations.

Where lakes are abundant, the surface deposits are either infertile, sandy, pitted glacial outwash, or boulder and clay morainic deposits. Poor soils and inadequate drainage dictate that the lake districts have limited agricultural potential. Consequently, the economy of these areas is geared to recreation and resort trade.

Waterfalls

The geology and climate of the Northern Highland produce a fine assortment of scenic waterfalls and rapids. The step, or stumbling block, necessary for a falls or rapids is formed when there are differences

Figure 3-5. A scenic lake in the Northern Highland. Trout Lake, Vilas County, Wisconsin. Wisconsin Department of Natural Resources photo.

in the resistance of adjacent rock units or within a single rock mass. Since running water preferentially removes weaker material, irregularities in the bedrock are soon emphasized.

Along the boundary between the western part of the Lake Superior Lowland and the Keweenawan basaltic ridges of northwestern Wisconsin, variation in adjacent rock types is provided by the Douglas fault (Fig. 2-3). The escarpment along this fault rises abruptly above the gently sloping glacial lake plain, and some of Wisconsin's highest falls occur where streams cross this elevation change on their way to Lake Superior. The rivers which enter Lake Superior leave the Northern Highland by the back door, since the major direction of drainage is south to Lake Michigan or the Mississippi River. These rivers drop about 350 feet in elevation between their headwaters and Lake Superior. The lake clay, sandstone, and conglomerate of the lowlands along Lake Superior are no match for the erosional power of the fast running rivers, and gorges up to 100 feet deep are cut in these soft materials. This erosion has further enhanced the differences in elevation along the fault.

Two Wisconsin parks are situated along the Douglas fault. Pattison State Park, 11 miles south of Superior, is located on the Black River, a tributary of the Nemadji. Here, the highest falls in Wisconsin leaps 165 feet over a basalt escarpment to cut a narrow canyon in the sandstone below. The Indians believed that the thundering voice of this falls was that of the Great Spirit Manitou; hence, the name Big Manitou Falls (Fig. 3-6). Little Manitou Falls is about a mile and a half upstream and it drops about 30 feet over a basaltic ridge. Pattison State Park also includes part of the Nickerson (Valderan) terminal moraine and several beach levels of Glacial Lake Duluth. About 12 miles northeast of Pattison, at Amnicon State Park, the Amnicon River cascades 20 feet over the fault escarpment into a quiet pine-topped sandstone canyon. To a geologist, the most important aspect of this park is that the Douglas fault is accessible in the gorge immediately below the falls (Fig. 2-15).

The water in the rivers of the north country is not your ordinary pallid stuff. It has sturdy character that is derived from iron-rich soil and swamp vegetation. The result is

Figure 3-6. Big Manitou Falls at Pattison State Park, Wisconsin plunges 165 feet over Middle Keweenawan basaltic lava that is uplifted along the Douglas fault. The gorge below the falls is carved in sandstone of the latest Precambrian to Middle Cambrian(?) Bayfield Group. Wisconsin Department of Natural Resources photo.

a rich, coppery brown fluid that develops a froth of white foam as it courses through the rapids. It resembles dark beer but it certainly doesn't taste like it.

Copper Falls State Park is another Wisconsin park that focuses on waterfalls. At Copper Falls, the Bad River plunges 30 feet over a resistant Keweenawan basaltic ridge into a narrow gorge (Fig. 3-2). About one-quarter mile downstream, Tyler Forks drops 30 feet over Brownstone Falls to join the Bad River. Three small falls connected by rapids are located along the Tyler Forks, upstream from Brownstone. Potato Falls, located in a county park south of Gurney (Iron County), Wisconsin, also owes its existence to resistant Keweenawan basalt. Upstream from the multilevel main falls is another cascade.

The Montreal River, which forms part of the Wisconsin-Michigan border, once bore an Indian name meaning "white falls river." A series of cascades begins at Gile Falls near Hurley, where the west fork of the Montreal flows over resistant rocks of the Penokee-Gogebic Range (Fig. 3-2). Just downstream from Hurley on the main Montreal is Peterson Falls. As the Montreal twists northwestward toward Lake Superior, it drops 115 feet over Keweenawan basaltic ledges at Saxon Falls. At this locality, the Lake Superior District Power Company has a power plant with public access to river level via a 150 step stairway. Below the falls, the Montreal enters a steep-sided, narrow gorge where it stumbles for two miles over a series of rapids and small falls. The lower end of this canyon is cut into Upper Keweenawan sandstone and conglomerate which erode to fingerlike spires. At the mouth of the canyon, the vista opens to a low landscape. Here, the Montreal slides slowly through soft maple and river birch, saving its strength for a last leap over sandstone ledges at Superior Falls, several hundred yards from Lake Superior. Another power station is located here to utilize the 127 foot drop of the river.

The Black River,[1] like the Montreal, rises in Wisconsin and cuts through the Penokee-Gogebic Range as it flows north to Lake Superior. Along a stretch of the Black from Bessemer, Michigan to Black River County Park on Lake Superior, there are numerous waterfalls that are accessible from the Black River Parkway. These include Chippewa, Algonquin, Great Conglomerate, Potawatomi, Gorge, Sandstone, and Rainbow falls. In addition to scenery, these falls provide an opportunity to examine Middle and Upper Keweenawan rocks (Fig. 3-2). Portage Lake lavas are exposed at Algonquin Falls, the Copper Harbor Conglomerate can be seen in the Gorge-Potawatomi Falls area, and the Nonesuch Shale is present at the mouth of the Black River.

More than a hundred waterfalls occur in the Northern Highlands part of Michigan, and some are particularly interesting because of their geologic settings. One of these is north of White Pine, where the Big Iron River races over resistant units of the Upper Precambrian Nonesuch Shale at Greenwood Falls. Several others are present on the Keweenaw Peninsula where streams plunge off resistant Keweenawan rocks in their rush to Lake Superior. Eagle River Falls is especially important because it provides an exposure of the contact between the Middle Keweenawan Portage Lake Lava Series and the Upper Keweenawan Copper Harbor Conglomerate.

Another interesting locality is Warner Creek Falls at Palmer, in the Marquette Range. The Negaunee Iron Formation is exposed here. Sturgeon Falls near Loretto (Dickinson County), Michigan provides the opportunity to examine the unconformity

1. There are at least five Black Rivers within the region covered in this book.

between Lower Precambrian gneiss and Middle Precambrian (Animikean) conglomerate and quartzite. Other northern Michigan falls of note include: Bond Falls (near Paulding, Ontonagon County), Silver River Falls (near L'Anse, Baraga County), Tioga Gorge Falls (Nestoria, Baraga County), and Canyon Falls (near Alberta, Baraga County).

Another group of falls is located in northeastern Wisconsin, near U.S. 141 in Marinette County. Among these are Dave's, Long Slide, and Twelvefoot falls. They, too, owe their existence to differences in the durability of various Precambrian rock bodies.

Monadnocks

A scattering of resistant hills, or monadnocks, rise above the general erosional surface of the Northern Highland (Fig. 2-4). Now quite modest in local relief and total elevation, these are the survivors of once lofty mountain ranges that dominated the landscape in the Precambrian. They endured several extensive periods of erosion, inundation by Paleozoic seas, burial by marine sediments, and the rigors of glaciation because they are composed of resistant Animikean (Middle Precambrian) quartzite.

The monadnocks rise from 300 to 600 feet above the surrounding terrain, but they seem much higher when the adjacent area is almost flat. Such is the case with Rib Mountain near Wausau, Wisconsin which stands 640 feet above the surrounding land and 763 feet above the nearby floodplain of the Wisconsin River (Fig. 3-7). Because of its imposing appearance, Rib Mountain, at an elevation of 1924 feet, was once thought to be the highest point in Wisconsin. However, more recent surveys passed the honor to

Figure 3-7. Rib Mountain in central Marathon County, Wisconsin is a monadnock of resistant Precambrian quartzite that rises 763 feet above the broad valley of the Wisconsin River. Wisconsin Department of Natural Resources photo.

Tim's Hill (elevation 1952.9 feet) in Price County. This work also established that Pearson Hill (Price County) and Sugarbush Hill (Forest County) were higher than Rib Mountain.

Although Rib Mountain is no longer first in elevation, this fact didn't curtail the winter ski activity here, nor is the mountain any less beautiful. The flanks are wooded and strewn with angular quartzite blocks, and the view from Rib Mountain State Park at the summit is spectacular. Mosinee and Hardwood hills, also in Marathon County, are smaller quartzite neighbors of Rib Mountain (Fig. 2-4). All escaped recent glaciation, since they lie in that rectangular projection of the Driftless Area that extends northward along the Wisconsin River into the Northern Highland (Fig. 2-28).

Flambeau Ridge is an elongate quartzite mound extending 3.5 miles along the northern Chippewa County line (Fig. 2-4). Its eastern end was breached by what may have been the preglacial Chippewa River. This gap, now high and dry, is partly filled with glacial drift, and it is similar to some that notch the Penokee-Gogebic Range to the north. Abandoned stream gorges of this type are called *wind gaps.* The wind gaps of the Northern Highland apparently formed when the Precambrian ridges had a cover of Paleozoic rocks. As the ancient rivers cut through these sediments, they encountered hidden obstacles beneath. Since the position of the river valley was already fixed, the running water began to erode the resistant rock. If it failed in this chore, a new channel was ultimately developed, but the gap remained.

The Barron Hills are northwest of Flambeau Ridge, and sprawl with a northeast trend from Barron County into parts of Rusk and Sawyer counties (Fig. 2-4). They are also known as the Blue Hills because of their bluish, hazy appearance. The highest relief in these quartzite hills is near Exeland in Sawyer County, where they rise 600 feet above the Chippewa River.

Smaller quartzite monadnocks in the Northern Highland include Powers Bluff in Wood County, and McCaslin and Thunder mountains in western Marinette County (Fig. 2-4). Resistant Animikean rocks also form the relatively high-standing iron ranges of northern Wisconsin and Michigan. These include the Penokee-Gogebic, Marquette, and Menominee ranges, which are discussed in a following section (Fig. 2-3).

Age relations of the resistant Precambrian quartzites of Wisconsin are not well-established. Some geologists suggest that the more northerly exposures are older rocks that were involved in an earlier deformation than those to the south.

Interstate Park

Wisconsin's oldest state park, located along the Dalles of the St. Croix River, was created in 1900 as a cooperative venture with Minnesota. Interstate Park is adjacent to the communities of St. Croix Falls (Polk County), Wisconsin and Taylors Falls (Chisago County), Minnesota, and it is now included as a unit of the Ice Age National Scientific Reserve (Figs. 2-36 and 2-37). When most people in our area speak of "the Dells" (a modernization of the French dalles), they are referring to the scenic gorge along the Wisconsin River in central Wisconsin. However, at Interstate Park, an equally scenic dalles was carved through glacial till, Upper Cambrian sandstone and flows of Keweenawan basaltic lava by torrents of glacial meltwater (Figs. 2-4 and 2-40).

The basalt along the river is well-jointed, so the cliffs are cut by vertical crevices that help to form blocky pinnacles. A number of these natural rock sculptures are named. One rocky shaft is known as the Devil's Chair, and a stern profile overlooking the gorge is now called Old Man of the Dalles

(Fig. 3-8). The Chippewa Indians considered this stony countenance to be Winneboujou, a god-hero of their legends. It was told how he watched over the Chippewa people from his lofty vantage point and several times caused this river to rise and destroy canoe flotillas of invading Sioux. It seems unthinkable that any white man would wish to replace Winneboujou with Old Man of the Dalles.

An earlier route of the St. Croix, which now rises near Solon Springs in Douglas County, Wisconsin, was located further west and it followed a different course to the Mississippi River. When glacial meltwater overflowed southward from Glacial Lake Duluth (ancestral Lake Superior), a new channel for the St. Croix was cut through glacial drift and Cambrian sandstone into the resistant basalt at Interstate Park. Remnants of the glacial drift are still present within the park and glacial striae on bedrock record earlier glacial advances.

The river's bedload of cobbles and gravel abraded the basaltic channel walls, and carved scores of cylindrical potholes (Fig. 3-9). Eddies in the fast moving water caused localized swirling of the bedload, which drilled down into solid rock as neatly as a carpenter's brace and bit. Many of these bedrock basins still contain cobble grinding tools within their interiors. The potholes range in size from about a foot in diameter, to great cisterns up to 25 feet across, with depths up to 80 feet. Present from river level to nearly 100 feet above the water, they are testimony to the erosive power of torrential glacial meltwater. To us, this is far more impressive than the fanciful names (Devil's Caldron, etc.) now assigned to these features.

Interstate Park is also an excellent place to view the unconformity between the Precambrian and the Paleozoic. On both sides of the St. Croix River, one can observe cobbles and

Figure 3-8. Winneboujou overlooks the Dalles of the St. Croix River at Interstate Park, Wisconsin-Minnesota. This gorge through Keweenawan lava flows was carved by the meltwaters of Glacial Lake Duluth that spilled southward down the Bois Brule (Figure 3-4) and into the St. Croix on its way to the Mississippi. Wisconsin Department of Natural Resources photo.

boulders of Keweenawan basalt surrounded by sands deposited during transgression of the Late Cambrian sea. This thin basal conglomerate, now a well-indurated rock, grades upward into sandstone. The unconformity and fossiliferous Upper Cambrian sandstone are most accessible on the canyon wall along U.S. 8 just west of Taylors Falls, Minnesota.

Chequamegon Bay and Apostle Islands National Lakeshore

Northeast of Ashland, the natural harbor of Chequamegon Bay opens to Lake Superior. Once utilized for the lake shipment of iron ore from the Penokee-Gogebic Range, this harbor is protected by a natural breakwater of sand that extends about 10 miles north toward the Bayfield Peninsula. Early explorers reported that the bar extended entirely across the bay in 1661, but storm waves later breached it in several places. A 4-mile stretch is now isolated from Chequamegon Point as Long Island (Fig. 3-2).

In spite of Figure 1-9, there is an Indian legend about the origin of this sand spit and the Apostle Islands beyond. According to Chippewa lore, the god Nanabozho was trying to capture the Great Beaver of Lake Superior. He threw handfuls of sand out into the lake to create this barrier and trap the beaver. Occasionally a handful would land farther out in the lake, and each of these "misses" formed one of the Apostle Islands.

Figure 3-9. A pothole carved by glacial meltwaters which flowed south from Glacial Lake Duluth; Interstate Park, Wisconsin-Minnesota. Wisconsin Department of Natural Resources photo.

The bay's name resulted from a French corruption of the Chippewa name for this escapade which meant "the place of the soft beaver dam."

The Nanabozho legend can only be partly correct since the Apostle Islands are not unconsolidated sand. They are composed of the same red-brown uppermost Precambrian to Middle Cambrian (?) sandstone that forms the Bayfield Peninsula and underlies the Pleistocene lake clay of the western part of the Lake Superior Lowland (Fig. 2-4). The Apostles are probably the remnants of hilltop areas between preglacial valleys which persisted as isolated islands during various stages of Glacial Lake Duluth and modern Lake Superior.

The erosional work of the lake created spectacular scenery along the shores of the Apostles and much of the Bayfield Peninsula. Precipitous cliffs carved by battering waves rise more than 60 feet above water level. A wealth of sea caves and arches accent the shore, and some of these are large enough for pleasure boats to enter. Pine and ash, highlighted by white birch, top the sandstone surfaces and provide a pleasing contrast with the brownish cliffs (Fig. 3-10). Interspersed are segments of sandy beach that soften the rocky shoreline.

A phenomenon referred to as "pillared rocks" occurs along the high shorelines at the peninsula's end and on some of the island margins. Here, symmetrical pillars with a sculptured appearance support a variety of arches. Devil's Island, in particular, is known for its interesting assortment of erosional forms. Wave action continues to modify the islands. Since the arrival of the white man, an island known as Steamboat was completely destroyed and another was cut into two parts.

Madeline Island, the largest and most populated of the Apostles, is served by a ferry from Bayfield, Wisconsin. On the is-

Figure 3-10. Sandstone of the latest Precambrian to Middle Cambrian(?) Bayfield Group along the Lake Superior shore in Apostle Islands National Lakeshore. Milwaukee Journal photo by George P. Koshollek, Jr.

land, there is a limited road network, an airfield, and Big Bay State Park. Because of this development, Madeline will not be included in the Apostle Islands National Lakeshore. Long Island, the separated end of Nanabozho's sand bar across Chequamegon Bay, will also be excluded. The National Park Service will manage the remainder of the island archipelago, and a strip of adjacent mainland shore; ultimately, 42,000 acres will be included.

Iron Ranges

The iron mining districts in the regions bordering Lake Superior are called ranges, since they are often linear topographic highs formed by resistant quartzite units associated with the iron formations. There are three iron ranges in northern Wisconsin and the Upper Peninsula. The Marquette Range is in Michigan, the Menominee Range is predominantly north of the Wisconsin-Michigan border with a small projection into northeast Wisconsin, and the Penokee-Gogebic is shared about equally by both states (Fig. 3-2). All of these ranges are visited on the Lake Superior field trip that is included in our companion volume.

The productive iron formations in the three ranges are Middle Precambrian (Animikean), and they are interbedded at several different horizons within a thick sequence of slate and quartzite, with lesser amounts of dolomite and conglomerate. *Iron formation* describes a siliceous, typically laminated or thin-bedded rock that contains more than 15% iron. These rocks are often reddish and are frequently banded with iron-rich and iron-poor layers. The type of iron mineral present varies in different districts, as does the interbedded material. A common iron formation contains layers of silica-rich rock, such as chert or jasper, alternating with layers that are rich in iron mineral such as hematite, magnetite, iron silicates, or siderite. Rocks of this type are called *taconite* or *jaspilite* and they usually contain 25% to 30% iron.

Iron formation apparently was deposited in shallow bays or basins. The iron was weathered from older, iron-rich rocks and carried to the depositional site by streams and ground water. Under favorable conditions, the iron precipitated in an alternating sequence with more siliceous beds that are believed to represent chemically precipitated silica. This cyclic deposition results in a unique type of sedimentary rock.

Animikean rocks were subjected to deformation during the Penokean orogeny at the end of Middle Precambrian time. As a result, they were folded, faulted, and intruded by igneous rock. This orogeny also metamorphosed sandstone, shale, and dolomite into quartzite, slate, and marble. Parts of the iron formation were weathered at various times during the Precambrian to form soft oxide ores of hematite and limonite. In some areas, a later metamorphic event recrystallized the iron formations and the weathered ores to create coarse-grained hard ores.

The earliest ores mined from the Lake Superior ranges contained up to 70% iron. Such deposits resulted from a concentration of iron by ground water as it dissolved and removed (*leached*) most of the siliceous material from the iron formation. These naturally enriched ores were sent directly from the mines to eastern steel mills. The first load was only a 6 barrel lot in 1852, but iron shipments boomed with the opening of the Soo Locks in 1855.

After World War II, the high grade, direct shipping ore in the Lake Superior region was depleted and it became necessary to use lower grade deposits. These leaner ores, such as taconite and jaspilite, presented several problems. In order to utilize them at a profit, improvements in both mining and processing technology were needed. The siliceous ores were much harder, so new procedures and equipment were developed. Once mined, it was necessary to artificially enrich or *beneficiate* this ore to increase the iron concentration to about 65%. The general steps involved in iron ore beneficiation are: crushing to release the iron-rich minerals, separation of these minerals by either magnetic or nonmagnetic means depending upon the nature of the iron minerals present, and a final agglomeration process that converts the iron-rich dust into easily handled pellets or briquets. This controlled product resulted in savings at the smelter, and a new era dawned for the venerable Lake Superior mining district.

The Marquette Range trends 30 miles west from Marquette, Michigan into east-central Baraga County, as a narrow belt of relatively resistant rocks. A part of this range extends southeast as a narrow trough to Republic, Michigan (Fig. 3-2). The structure of the Marquette Range is synclinal although it is complicated by many faults, smaller folds, and igneous intrusions. This was the first of the Lake Superior iron mining districts, and mining which began in 1848 continues today. Ore is presently produced from the Negaunee Iron Formation, although the Bijiki

Iron Formation was also productive in the past. This range is a rugged highland area of lakes and swamps, that forms a drainage divide between Lake Superior and Lake Michigan. Along the range are the historic mining towns of Marquette, Palmer, Negaunee, Ishpeming, Champion, and Michigamme.

The main part of the Menominee Range is located in Iron and Dickinson counties, Michigan, with a short extension into Florence County, Wisconsin (Fig. 3-2). The structure of this range is a northwesterly trending syncline that is complicated by smaller folds and faults. The productive horizon in Dickinson County is the Vulcan Iron Formation, which has an average iron content of 30%. Enrichment of this lean ore occurred in narrow, synclinal troughs where the movement of ground water leached silica and concentrated iron. To the northwest, in the Florence and Crystal Falls-Iron River districts, the productive horizon is the younger Riverton Iron Formation.

Mines were opened along the Menominee Range between 1872 and 1880. Ore was shipped by wagon to Menominee, Michigan for Great Lakes transport until the Chicago and Northwestern Railroad provided a shorter route to Escanaba, Michigan in 1877. Mining is still underway in the Crystal Falls district and at Felch, north of Iron Mountain.

The Penokee-Gogebic Range extends 80 miles northeast from Namekagon Lake in Bayfield County, Wisconsin to Lake Gogebic in Michigan (Fig. 3-2). This range is a steeply north-dipping Middle Precambrian sequence of quartzite, slate, and iron formation that is faulted and cut by igneous intrusions. The resistant rocks at the base of this succession form a conspicuous ridge up to a mile wide. North of this ridge is a lowland developed on less resistant slate at the top of the sedimentary sequence. The principal mining communities, the Soo Railroad, Wisconsin Highway 77 between Mellen and Hurley, and U.S. 2 from Hurley, Wisconsin to Wakefield, Michigan are located in this lowland.

The Penokee-Gogebic Range is breached by many northward-flowing streams. There are 7 along the 30-mile stretch from the Bad River at Mellen, Wisconsin to the Montreal River at Hurley, Wisconsin, and another 5 cut the Michigan part of the range. Most of these streams utilize passes through the Penokee-Gogebic that were formed when ancient streams flowing in valleys of Paleozoic sedimentary rock gradually cut down into the buried Precambrian ridge beneath. Not all of these streams were successful in eroding through the resistant ridge, and some courses were abandoned. The Paleozoic sediments were subsequently removed, and in postglacial times, new streams found it relatively easy to remove Pleistocene fill and reoccupy the old valleys. Such features are called *water gaps.* Those cuts that represent stream courses that were abandoned form wind gaps, similar to the one we previously described on Flambeau Ridge in northern Chippewa County, Wisconsin. The major water gaps provide convenient north-south routes for highways and railroads.

The western end of the range, now known as the Penokee, was more or less named by geological surveyor Charles Whittlesey in 1849. He named the range Pewabic, an Indian word for iron, but the printer misread the name and the result was Penokie. With subsequent usage, this became Penokee. However, the surveyor was ultimately successful with another name, and the highest point on the range (an elevation of 1,872 feet) is now called Mt. Whittlesey.

There is a tale told in the sister cities of

Hurley, Wisconsin and Ironwood, Michigan about a pine tree which blew over in 1884 to reveal the presence of rich iron ore. This, and the virgin pine forests, started a rush that brought thousands of miners and lumberjacks to these booming towns. Hurley soon acquired a tough and rowdy reputation. It later became a rendezvous for crime figures, adding bawdiness, gambling and violence to the list during the 1920s and 30s. Ironwood developed more peacefully as the business center for the two cities. More recently, ski developments along the range converted both communities into year-round recreational centers.

The Penokee-Gogebic mines were all underground because they followed the steeply dipping Ironwood Iron Formation. During the shipping season, ore was loaded into railroad cars and dispatched to Ashland, Wisconsin where it was loaded on Great Lakes vessels. The Montreal iron mine near Hurley produced more than a million tons of ore annually, from depths greater than 4,000 feet below the surface. This mine produced more ore than any other Wisconsin iron mine until its closing in 1962. In spite of such an impressive contribution, most of the total ore production (255 million tons) was from the Michigan (or Gogebic) part of the range. The Cary mine at Hurley was Wisconsin's last operating mine on the range, when it closed in 1965.

Copper Country

The Keweenaw Peninsula of Michigan is a narrow, rocky spine that projects 65 miles into Lake Superior (Fig. 3-2). This is Copper Country, where deposits of native copper occur in greater abundance than anywhere else in the world. The mineral bonanza here was discovered, mined, and utilized nearly 4,000 years ago by Indians. These Copper Culture Indians, who used metal rather than stone or bone for many of their implements, vanished long before Europeans came upon the scene. However, their mine pits are still visible in this area and on Isle Royale. Although later Indians didn't mine or use the copper, pieces were collected from glacial drift and traded widely.

Finds of float copper in this area were reported by Jesuit explorers in the 1600s, but little interest developed until 1841 when Douglass Houghton, the first state geologist of Michigan, reported that these copper deposits were economically important. The boom was on, and mining began at Copper Harbor at the tip of the Keweenaw Peninsula in 1844. Towns sprang up overnight to accommodate the rapid inflow of settlers. Discoveries of rich lodes continued for 40 years and the Keweenaw was king of copper. Unfortunately, Douglass Houghton didn't live to see his prediction fulfilled. He drowned along the stormy shores of the peninsula while rushing to complete field work on the geology of this area.

The richest and most accessible ores were eventually exhausted, and production began to drop off when underground mining to greater depths became increasingly expensive. After a time, leadership in United States copper production shifted to Butte, Montana and presently to Bingham Canyon, Utah. Several mines struggled on into the mid-1960s, and depths of more than 6,000 feet were reached. No mines are operating now (1976) in the Keweenaw, although copper is produced nearby at the White Pine mine. However, exploration and research on extraction techniques continues and some mines may open again. While the miners await this happy day, they can take pride in recalling that Keweenaw copper came along at the right time to accelerate the electrification of the nation and it helped the United States through World Wars I and II.

The native copper deposits occur in the Middle Keweenawan Portage Lake Series (Fig. 3-2). This formation, with at least 400 individual lava flows and some interbedded conglomerate, dips an average of 40 degrees northwest toward the center of the Lake Superior syncline. Although the relics of mining activity extend along the outcrop belt of the Portage Lake for 90 miles, almost all of the copper production came from a 30 mile stretch from Painesville to Eagle River (Fig. 3-2). Of this production, about half came from one conglomerate interbed (Calumet Conglomerate).

The copper was formed by hot, mineral-bearing waters that moved upward from depth, either during or soon after accumulation of this thick sequence of lavas in the Lake Superior syncline. The circulating waters passed through porous rocks or cracks within impervious rocks, and deposited copper in any available opening. These included cracks or fissures, vesicles within lavas, fragmental surfaces at the top of some flows, and voids between the pebbles and sand grains of the conglomerates. The cumulative production of copper from veins or fissures was not great, but the largest individual masses were found in such deposits. The Ontonagon Boulder displayed at the U.S. National Museum consists of two tons of copper that formed in a fissure about a billion years before it was collected and deposited as an erratic by glacial ice.

In addition to the copper-bearing rocks of the Portage Lake Series, the northern tip of the Keweenaw Peninsula provides exposures of the Upper Keweenawan Copper Harbor Conglomerate (Fig. 3-2). The younger Upper Keweenawan formations (Nonesuch and Freda) are present along the northwest side of the lower half of the peninsula, but they rarely outcrop (Fig. 3-2). On the opposite side of the Keweenaw is a lowland developed on the latest Precambrian to Middle Cambrian (?) Jacobsville Sandstone. This lake-studded lowland lies southeast of the Keweenaw fault which extends from just east of Houghton-Hancock, to immediately northwest of lakes Gratiot and La Belle, and into Bete Grise Bay (Fig. 3-2). This fault brings resistant lavas of the Portage Lake Series up and over the Jacobsville. Although rarely exposed, it is well-displayed at the Natural Wall near Laurium (Fig. 2-14).

The highland area of lavas northwest of the Keweenaw fault is called the Copper Range. The surface of this feature consists of small, linear ridges and valleys formed by outcroppings of Portage Lake units in a trend that roughly parallels the shoreline. More resistant rocks form steep, southeast facing ridges that dip northwest into the Lake Superior syncline. Some of the softer, valley-forming units contain lakes, such as Medora, Meadow, Upson, Bailey, and Lake Fanny Hooe at Fort Wilkins State Park.

Picturesque harbors along the northwest shore of the peninsula also reflect bedrock control of topography. The waters of Lake Superior enter narrow inlets through the tilted edges of resistant rock layers to broaden into linear bays developed in more easily eroded rock units. These harbor names suggest the character and resources of the Upper Peninsula—Copper, Eagle, and Agate. The latter is a reminder of the fine mineral collecting along the Lake Superior shore of the Keweenaw Peninsula.

Porcupine Mountains

The Porcupine Mountains of Michigan are a state park on the south shore of Lake Superior. Like much of the Keweenaw Peninsula 60 miles to the northeast, the bedrock here includes the Keweenawan Portage Lake Lava Series, Copper Harbor Conglomerate, Nonesuch Shale and Freda

Sandstone (Fig. 3-2). The youngest two formations (Nonesuch and Freda) are present only at the western edge of the park along the picturesque gorge of the Presque Isle River, and immediately east of the park along the Big Iron River. The Nonesuch is especially well exposed along the Big Iron at Greenwood Falls.

Porcupine Mountains State Park includes a stretch of Lake Superior shoreline that provides exposures of the Copper Harbor Conglomerate. In this area the Copper Harbor is more than 5,000 feet thick and consists primarily of sandstone with minor amounts of conglomeratic material. A thick series of lava flows, the Lakeshore Traps, are a part of this unit, and they are accessible at the lofty overlook of Lake of the Clouds at the end of Highway M-107 (Fig. 3-11). The flows at this location are amygdaloidal lavas.

The Porcupine Mountains are high-stand-

Figure 3-11. Lake of the Clouds in Porcupine Mountains State Park, Michigan. Amygdaloidal basaltic lavas of the Upper Keweenawan Lake Shore Traps dip northwest off the Porcupine Mountains anticline toward the center of the Lake Superior syncline. Several individual flows are defined by crude bedding planes. Lake of the Clouds occupies a glacially modified river valley. Michigan Tourist Council photo.

ing because they are on the north-dipping limb of a large, asymmetrical anticlinal uplift (Fig. 3-2). Information on the other limb of the anticline is based on subsurface studies related to copper exploration south of the park boundary. This work suggests that southerly dips are very steep and beds may locally overturn. The high, crestal part of the anticline brings up rocks as old as the Portage Lake Series (Fig. 3-2). This area, a rugged, tree-covered wilderness, lies north of Lake of the Clouds. The highest point in the park is Government Peak at an elevation of 1,863 feet; about 1,260 feet above Lake Superior.

Since the Freda Sandstone is involved in the anticlinal folding, we know that the Porcupines formed after Late Keweenawan time. Unfortunately, this does not help put an upper limit on the age of folding. It seems reasonable to assume that the Porcupine Mountains anticline and the adjacent Iron River syncline are both related to the compressional forces that resulted in the Keweenaw fault farther to the south (Fig. 3-2). Since we dated the Keweenaw fault in Chapter 2 as latest Precambrian to Early Cambrian, we conclude that this is also the time of the Porcupine Mountains folding.

Once folded, the Porcupine Mountains were subjected to a long period of erosion and ultimately, glaciation. Glacial grooves and chatter marks on bedrock surfaces show that the latest Wisconsinan ice advance moved from east to west through the park.

Isle Royale National Park

Isle Royale, the largest island in Lake Superior, floats in solitude about 45 miles away from the rest of northern Michigan (Fig. 3-2). Closer to Canada than the United States, the international boundary makes a pronounced jog to accommodate it. The name for this 45 mile long and 7 mile wide island was provided by French voyageurs about 1670 in honor of their royalty, King Louis XIV. Today the island is a national park whose boundaries encompass three times more water than land. Entry to the park is provided by water and air, while island travel is limited to boats or hiking on 120 miles of trails. Passenger boats from Houghton and Copper Harbor, Michigan, and Grand Portage, Minnesota operate during the summer months. It is advisable to plan a trip to Isle Royale well in advance, because of limited facilities and sharply increased visitor interest in this beautiful island.

Rising more than 700 feet above the present lake surface, Isle Royale has a washboard topography of northeast-southwest trending ridges and valleys. Although several of these ridges extend the entire length of the island, the pattern is partially interrupted by a cross-cutting system of fractures which also influences topography and drainage. The ragged, embayed pattern of the shoreline, like the washboard topography, reflects individual variation in the erosional resistance of Middle Keweenawan Portage Lake lava flows. Although 10,000 feet of Portage Lake lavas with minor interbeds of conglomerate and sandstone are exposed on the island, this is only about one-quarter of the total thickness of the formation.

Overlying the Portage Lake on the extreme southwestern end of the island is the Copper Harbor Conglomerate (Fig. 3-2). The pebbles and sand in this formation were derived from an older volcanic area somewhere to the northwest. From this source, sediments were carried southeasterly by rivers that deposited them in this area as alluvial fans and flood plain deposits.

The center of the Lake Superior syncline is midway between the Upper Peninsula shore and Isle Royale (Fig. 3-2). Immediately north of the island is the steeply dipping Isle Royale fault along which Keweenawan rocks of the park moved upward and northward. This fault accounts for the existence of this island on the north rim of the Lake Superior syncline (Fig. 2-10). The rocks on the island dip 15 degrees southeast under the waters of Lake Superior only to reappear on the south rim of the syncline along the Keweenaw Peninsula, where they dip northwestward at 40 degrees (Fig. 3-12).

Copper occurrences similar to those of the Keweenaw Peninsula are present on Isle Royale. The Indians of the Copper Culture were the first copper miners on the island, and they dug shallow pits with stone tools to quarry the mineral-bearing rock. Once extracted, the rock was heated in a fire before giving it a shock treatment of icy Lake Superior water. This process fractured the ore into more workable sizes. The copper was liberated with pounding stones; many of

Figure 3-12. Middle Keweenawan basaltic lavas at Isle Royale National Park dip gently southeast toward the center of the Lake Superior syncline. Bedding defines individual flows. The bedrock is rounded and polished by glacial ice. Offshore islands are common along this ragged shoreline. Michigan Tourist Council photo.

which were imported because of their special hardness. Minong, meaning "a good place to get copper," was the Indian name for the island.

Prospectors scoured the island in the mid-1840s, when the island officially became part of the United States. Commercial mining was attempted in the 1870s, but efforts ended by 1899 because the ore deposits were not plentiful nor continuous enough.

Vulnerable in its exposed position, Isle Royale was invaded by ice during all four Pleistocene glaciations. However, each obliterated the evidence left by the previous ice advance. As a result, the only glacial features preserved on the island date from the final Wisconsinan (Valderan) advance about 11,000 years ago. The direction of the Valderan ice invasion approximately paralleled the northeast-southwest strike of the bedrock for about two-thirds of the island's length. Glacial quarrying accented the ridge and valley topography, and abrasion was so effective that striated, polished, and rounded bedrock surfaces are common (Fig. 3-12).

On the western end of the island, the Valderan ice movement was at an angle to the ridge and valley system. Here, glacial drift is more abundant and it tends to subdue the landscape. The ice paused west of Siskiwit Lake as it receded, and deposited morainic debris.

As the Pleistocene drew to a close, Isle Royale was partially submerged by the meltwater of an ancestral stage of Lake Superior. Slowly, the ice-liberated land rebounded, and eastern drainage outlets opened, and lowered the water level. Several stages of island emergence are documented by abandoned shorelines well above the present level of Lake Superior. Features along these ancient shorelines include sea caves, arches, and remnants of sand or cobble beaches, some up to 200 feet above the present shoreline. This area is still rebounding, so today's beaches will also be abandoned in the future.

This land of rocky, often barren ridges, with intervening lowlands occupied by lakes and swamps, is exciting on its geologic merits alone. However, equally intriguing is the fauna of Isle Royale which includes moose and timber wolves.

Chapter 4

CENTRAL PLAIN

GENERAL CHARACTER

In contrast to the Northern Highland, the other three physical provinces in our area are parts of the Central Lowlands of North America (Figs. 2-2 and 3-1). This is a region of moderate topography developed on Paleozoic rocks, which was greatly modified in the northern part by Pleistocene glaciation. Within our area, the Central Lowlands is divided into the Central Plain, the Eastern Ridges and Lowlands, and the Western Uplands. Each of these physical provinces is in contact with the Northern Highland and with each other (Fig. 3-1).

The Paleozoic rocks in these provinces are the retreating erosional edges of marine sediments that once extended across the Northern Highland. Late Paleozoic uplift of the Wisconsin dome, and subsequent erosion, resulted in a belted bedrock pattern with progressively younger rocks exposed farther away from the center of uplift (Fig. 2-4 and 4-1). The amount of dip, the thickness, and the erosional resistance of individual formations determines the width of outcropping rock units. Whether a bedrock unit forms a lowland or a *cuesta* (a gently dipping, linear ridge of rock) depends upon the relative erosional resistance of adjacent rock units (Fig. 4-1).

The Central Plain is our smallest province and it is entirely within Wisconsin. It is a crescent-shaped lowland with a maximum dimension of 55 miles north to south and about 165 miles east to west. This province also has a relatively simple geologic history. It is largely developed on Cambrian sandstone, and it lies between the Precambrian rock of the Northern Highland and the first cuesta of more resistant younger Cambrian and Ordovician sandstone and dolomite

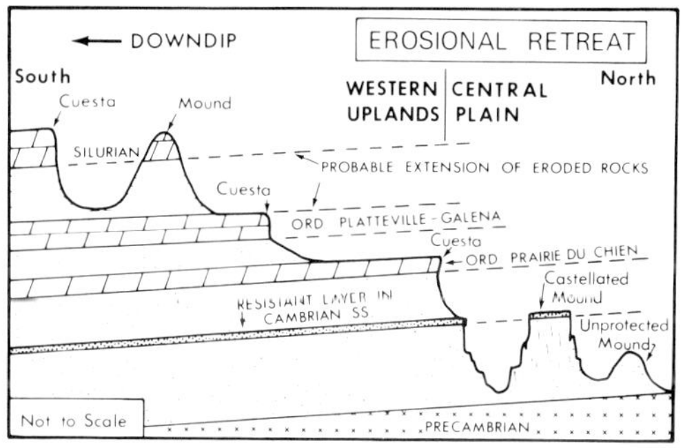

Figure 4-1. Diagrammatic cross section of the Western Uplands and Central Plain illustrating formation of cuestas and mounds by downdip erosion of gently dipping sedimentary strata with varied resistance to weathering. A castellated Cambrian mound like those of the unglaciated part of the Central Plain is shown in the lower right. To the far right is a later erosional stage of such a mound when the protective cap rock is gone. The resistant dolomite cuestas of the Ordovician Prairie du Chien, the Platteville-Galena, and the Silurian are indicated. The isolated mound capped by resistant Silurian dolomite in the upper left is similar to the mounds that are present in the Western Uplands of southwestern Wisconsin and northwestern Illinois. These are called outliers, since they are out from the main outcrop area.

(Fig. 4-2). The surface of the plain slopes gently southward from the Northern Highland at about four feet per mile, until it reaches the confining cuestas at the edges of the Western Uplands and the Eastern Ridges and Lowlands. These cuestas rise 150 to 350 feet above the Central Plain (Fig. 4-1).

The topography of the Central Plain is flat or slightly rolling, with little relief except for glacial deposits to the east and northwest, and isolated sandstone mounds in the Driftless Area. Despite the limited distribution of these steep-sided mounds, they are the most unique feature of the Central Plain. In fact, a traveler driving between Wisconsin Dells and Tomah might believe that he had arrived in the butte and mesa country of the western United States.

The sandy soils of this region are generally poorly suited for agriculture, although they support a mixed vegetation that includes important amounts of timber. Oak and jack pine barrens are interspersed with white and red pine, representative of more northern forests. Where the relatively infertile soil is underlain by lake clays, tamarack bogs flourish.

Geology influenced settlement along the northern edge of the Central Plain. Here, rivers which flow from the Northern Highland leave resistant granite and gneiss and easily deepen their channels in the softer sandstone (Fig. 2-4). Where this occurs, a line of cascades or low rapids is developed. This fall line marks the upper limit of river navigation, and it is similar to the one that influenced pioneer settlement along the historic fall line of the eastern United States. In Wisconsin, the falls on the larger rivers were ideal for the development of sawmills to serve the early lumbering industry. Wisconsin Rapids, Stevens Point, Chippewa Falls,

Figure 4-2. The Cambrian sandstone cuesta at the border between the Central Plain and Western Uplands; south of Mauston (Juneau County), Wisconsin. Immediately north of the cuesta is a small mound with a cap of resistant sandstone. The low hill on the right side of the photo has lost its protective cap rock.

Waupaca, Nekoosa and Port Edwards are cities that developed along the fall line between the Northern Highland and the Central Plain.

Much of the region is sparsely settled and small farms alternate with large tracts of woodland and swamp. A significant part of the land is in school, county, state and federal forest and wildlife reserves. Agriculture is more important in the northwestern part where older glacial deposits provide suitable soils. The eastern part is underlain by young glacial deposits with numerous lakes that support an active tourist industry. The scenic Dells of the Wisconsin River, the primary tourist attraction in the state, is located in the south-central part of this province.

BEDROCK AND STRUCTURE

Late Cambrian seas washed over the Precambrian erosional surface to deposit more than 1,000 feet of sandstone, with minor amounts of interbedded dolomite. Ripple marks and cross-bedding in these rocks testify to nearshore, shallow water deposition. A rather limited group of inhabitants populated this ancient coastal region. Trilobites, extinct segmented scavengers, and simple bivalves (brachiopods) lived and died in these shallow seas. Other tenants left only networks of burrows and trails.

The Wonewoc Sandstone, 100 to 400 feet thick, is a mound and cliff former in the Central Plain. It is this formation that enhances the scenery of Wisconsin Dells, and at state parks that include Mill Bluff, Mirror Lake, Roche A Cri, and Rocky Arbor (Figs. 2-17 and 2-38). Well-cemented sandstones near the top of the Wonewoc are also important to the survival of the solitary mounds (Fig. 4-3). These more resistant layers act as a protective cap against the forces of erosion (Fig. 4-1).

A few low, rounded knobs in this region are not sandstone hills. Should you attempt to chip off a sample for a closer look, a rebounding hammer and a hearty clang will announce Animikean quartzite. Necedah Mound (Juneau County), Middle Mound at Babcock (Wood County), and Hamilton Mound (Adams County) are small quartzite hills that protrude through the Cambrian sandstone. These features once stood as resistant knobs on the subdued Precambrian erosion surface before burial by Cambrian sediments. Erosion is now working to restore them to their former glory. The rugged Baraboo Range of south-central Wisconsin is a similar feature within the Western Uplands, and it is discussed in detail in Chapter 5.

Other types of Precambrian rocks are locally present within the Central Plain. Granite and gneiss are exposed along major drainages in the vicinity of Eau Claire, and along the Black River between Neillsville and Black River Falls (Clark and Jackson counties) (Fig. 2-4). An Animikean taconite is also being strip-mined near Black River Falls. Granite and rhyolite outcrop near Redgranite and Spring Lake in Waushara County, in the vicinity of Berlin in Green Lake County, and at Montello in Marquette County. The red Montello granite made its claim to national fame as a building stone for the tomb of Ulysses S. Grant in New York City.

GLACIATION

The topography of the Central Plain reflects differences in Pleistocene history within the area. In that part of the Central Plain that is assigned to the Driftless Area, steep-sided sandstone mesas and buttes (often called *castellated mounds*) are distinctive landforms (Fig. 2-28). It is likely that similar mounds existed throughout the Central

Figure 4-3. Roche A Cri State Park (Adams County), Wisconsin. This mound rises precipitously from the flat, sandy floor of Glacial Lake Wisconsin. Photo courtesy of James Newton.

Plain in preglacial time, although now they are only concentrated in parts of Jackson, Monroe, Wood, Portage, Juneau and Adams counties.

The nearly flat surface of much of the landscape in the unglaciated part of this province is due to uniform erosion of the soft, almost horizontally-bedded sandstone, and to deposition in Glacial Lake Wisconsin, which filled some irregularities (Fig. 2-28). This 1,825 square mile impoundment in the northeast corner of the Driftless Area formed between Wisconsinan ice margins on the north and east, and areas of higher elevation to the south and west. Finely ground rock particles and detached blocks of stagnant ice were washed into this lake and deposited as flat-lying beds of sand and clay, that also contain a few erratic boulders which melted free of the floating icebergs.

Many of the Cambrian sandstone mounds and buttes were once islands in Glacial Lake Wisconsin (Fig. 4-3). A search around the base of many of these mounds will disclose evidence of this abandoned shoreline. Although these steep hills are spectacular features today, they must have been even more picturesque as rocky ships at sea in this great inland lake.

In contrast to the unglaciated part of this province, the remainder of the region, to the northwest and east, lacks castellated mounds and has a rolling cover of ground moraine, sandy outwash, and clay-rich lake deposits. Each of these glaciated areas has a somewhat different character. In the northwestern corner of the Central Plain, north of a line from Wisconsin Rapids to a few miles south of Eau Claire, the bedrock is largely covered by an older glacial drift than that which blankets the eastern glaciated area. This thin, maturely weathered glacial cover was deposited during the earliest Wisconsinan (Altonian) glaciation more than 30,000 years ago and the result is a subdued, rolling landscape (Fig. 2-28).

The eastern part of the Central Plain was covered by younger Wisconsinan ice advancing from the northeast about 22,000 to 12,500 years ago. The terminal moraine at the eastern border of the Driftless Area marks the maximum westward extent of the Green Bay lobe during this Woodfordian advance (Fig. 2-28). It can be traced across the Central Plain as a conspicuous feature from near Baraboo to Wisconsin Dells, and immediately east of Hancock and Stevens Point. The drift behind the morainal ridge system is thick, sandy and similar in topographic expression to that found in the Northern Highland, the northern part of the Western Uplands, and the Eastern Ridges and Lowlands provinces.

A broad, almost flat outwash apron was built immediately west of the terminal moraine. It extends westward more than 15 miles and is now the locale of extensive irrigation agriculture. The waters of Glacial Lake Wisconsin lapped up onto the western margin of this outwash plain. Considered part of the Driftless Area, the plain extends into the Northern Highland (Fig. 2-28).

As the Green Bay lobe of the Woodfordian ice retreated, a series of well-developed recessional moraines formed across Waupaca County. A forerunner of Glacial Lake Oshkosh also developed during this time. This lake was impounded by the retreating ice front and surrounding areas of higher morainal deposits (Fig. 2-28). It probably first overflowed southward via the Rock River, but later it drained into the Wisconsin River.

Glacial Lake Oshkosh formed later, during the retreat of the Green Bay lobe of Valderan ice. At this time, it occupied most of Winnebago County and parts of adjacent counties, and extended north into present-day Green Bay. Impoundment was provided by the terminal and recessional moraines of the Valderan Green Bay lobe and by the retreating ice front (Fig. 2-28). Only the northwestern part of Glacial Lake Oshkosh was within the area of the Central Plain, while the remainder extended into the Eastern Ridges and Lowlands.

DRAINAGE

The Central Plain is crossed by a series of streams which flow off the Northern Highland. Of these, the Wisconsin River is the largest, and its course across the sandstone lowland is characterized by a gentle grade and few tributaries. Although the river flows in a shallow trench near Wisconsin Rapids, it occupies a broad valley south of this point to immediately north of Wisconsin Dells. Along the Juneau and Adams county line, the Wisconsin River has been impounded by power dams to form the large Castle Rock and Petenwell flowages.

During the Pleistocene, the Wisconsin River north of Portage was diverted westward by the advancing Woodfordian ice front (Figs. 2-28 and 4-4). The valleys of its eastern tributaries, such as the Roche A Cri and others between Stevens Point and Portage, record the passing of large volumes of meltwater. These streams now head on the flanks of the terminal moraine which assumed the role of a local drainage divide when the glaciers receded. The Yellow River, entering the Wisconsin near Necedah, carried large quantities of meltwater from the ice front to the north. In contrast, the meandering Lemonweir, which enters the Wisconsin near Mauston, flows on sediments of Glacial Lake Wisconsin and has no other glacial history.

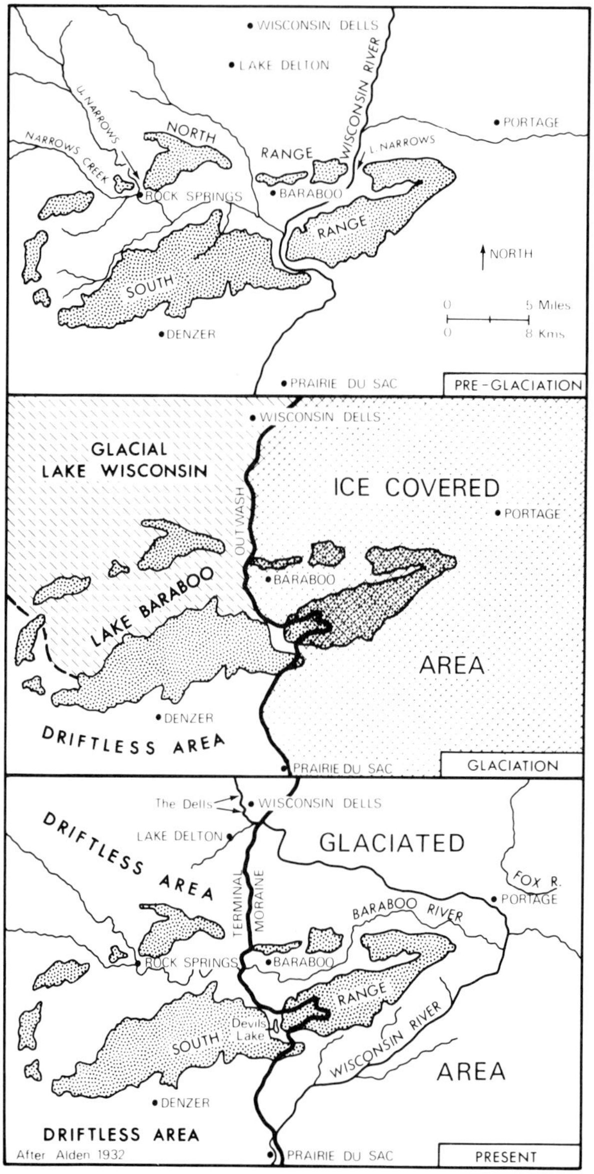

Figure 4-4. Pleistocene history of the Wisconsin Dells-Baraboo region. The upper map is an interpretation of the drainage in relation to exposures of Precambrian Baraboo Quartzite (stippled) in this area *prior to Woodfordian* glaciation. The South Range is higher and wider than the North Range and more extensive exposures of quartzite project through flat-lying Paleozoic rocks. The drainage, including an ancestral Wisconsin River flowing south, cuts through the quartzite in several places such as the Upper and Lower Narrows and Devils Lake gap.

The middle map shows *Woodfordian glaciation,* development of glacial lakes Baraboo and Wisconsin, and the Driftless Area. The terminal moraine loops around to close off the ends of the Devils Lake gap.

The lower map shows the *present-day* Wisconsin River flowing south through the Dells and the terminal moraine, and around the Baraboo syncline. Development of a passage through the moraine initiated the draining of Glacial Lake Wisconsin. The rush of meltwaters that followed carved through Cambrian sandstone to form the Dells. Portage, between the Fox and Wisconsin rivers, is on the upper right corner of the map. The Baraboo River now enters the syncline through the Upper Narrows and exits through the Lower Narrows. Devils Lake is within the gap cut by the preglacial Wisconsin River that is now plugged at each end by the Woodfordian terminal moraine.

A subcontinental divide between drainage to the Atlantic Ocean (by way of the Fox River to the Great Lakes and the St. Lawrence) and to the Gulf of Mexico (by way of the Wisconsin and Mississippi rivers) exists at Portage. Here, both the Wisconsin and the Fox rivers flow across a flat, swampy plain within 1.5 miles of each other (Fig. 4-4). Indians and early explorers, including Marquette and Joliet on their historic journey to the Mississippi, portaged their canoes across this stretch. During flood stages, the Wisconsin River would inundate the marshy divide, since its normal level is 3 feet higher than that of the Fox. At such times, canoes could be floated between the two drainages. A levee now protects the city of Portage from floods of this magnitude.

Construction of a canal to connect the Fox and Wisconsin at Portage began in 1838, and continued in spite of various setbacks for almost 40 years. Although a small steamer made an epic passage from the Mississippi River to the Great Lakes during the high water period of 1856, the canal was not open to larger boats until 1876. After a few years of operation, the cost of the constant dredging required to maintain a channel for large vessels in the shallow Wisconsin proved prohibitive. This problem, and increased competition from an improved railroad network, caused river shipments to wane. The locks at Portage were finally closed in 1951, and the abandoned canal is now a stagnant reminder of the past.

The Wisconsin is the master stream crossing the Central Plain, but other important drainages also traverse the region. To the west, the Black and Chippewa rivers flow off the Northern Highland to the Mississippi. Both cut through the Cambrian sandstone to expose Precambrian rocks within the Central Plain (Fig. 2-4). Where this occurs, falls and rapids are developed some distance south of the main Precambrian-Cambrian fall line. At Eau Claire and Black River Falls, as along the primary fall line, water storage facilities for hydroelectric plants were developed at rapids.

There are several historic aspects of the Black that are worthy of mention. A stretch of the Black River southwest of Hatfield apparently served as an outlet channel for Glacial Lake Wisconsin until the receding ice margins allowed for a change. As the ice retreated, the lake waters escaped through the terminal moraine to the southeast and carved the Dells of the Wisconsin (Fig. 4-4). In more recent times, the Black flooded in 1911 and wiped out 42 acres of the main business district of Black River Falls.

Among those rivers rising in the Northern Highland and draining across the eastern part of the Central Plain into Green Bay and Lake Michigan, are the Wolf-Fox system and the Oconto. In spite of these river systems, much of the eastern part of this province is poorly drained because of the irregularity of Late Wisconsinan glacial deposits. Consequently the area is occupied by numerous lakes, swamps and marshes that are connected by low gradient streams that wander aimlessly between higher, rolling woodlands. The most distinctive lake district in this province is confined to the Woodfordian morainal systems in Marquette, Waushara, Waupaca and eastern Portage counties. Most of these beautiful, small lakes were kettles, but some fill other types of depositional irregularities. This lake district, like similar ones in the Northern Highland, is an important tourist area.

The upper Fox River has a gentle gradient of less than 6 inches per mile as it traverses sediments deposited in an ancestor of Glacial Lake Oshkosh. Two linear water bodies mark the river's course as it widens, and the current diminishes at the shallow impound-

Figure 4-5. A tour boat plying the Upper Dells of the Wisconsin River. This narrow, scenic gorge was carved through Upper Cambrian Wonewoc Sandstone by meltwaters spilling out of Glacial Lake Wisconsin. Photo courtesy of Wisconsin Dells Chamber of Commerce.

ments of Buffalo and Puckaway lakes. Lake Poygan, a similar shallow enlargement of the Wolf River, and Shawano Lake, nearer the headwaters of the Wolf, are also developed on glacial lake sediments (Fig. 2-28). All of these lakes may have originally been irregularities on the floors of ancient glacial lakes. Green Lake, scooped to a depth of about 230 feet, is the deepest inland Wisconsin lake. It was carved from soft Cambrian sandstone along the edge of the Ordovician Prairie du Chien escarpment by an efficiently eroding part of the ice sheet.

The northwestern part of the Central Plain is blanketed by older glacial drift (Altonian) and the drainage systems are now well-developed (Fig. 2-28). Consequently, this region lacks the lakes and swamps associated with younger glacial deposits to the east.

One might expect the central, unglaciated part of the Central Plain to be well-drained, because of the porous sandstone bedrock. However, the largest swamp in Wisconsin, with an area greater than two Milwaukee Counties, is located within the triangular area roughly defined by Wisconsin Rapids, Tomah and Portage. Within these borders are the Necedah National Wildlife Refuge, Sandhill Wildlife Area, the Central Wisconsin Conservation Area and parts of the Black River State Forest. The Necedah Refuge and parts of the Central Wisconsin Conservation Area are the sites of former large-scale attempts to farm poorly-drained, marginal lands. Today, they are a recreational paradise for the hiker, camper, hunter and cross country skier.

Most of this swamp was once the site of Glacial Lake Wisconsin, and the surface sediments are often sandy remnants of beaches, bars and dunes associated with this ancient lake (Fig. 2-28). This is the sand country beloved and immortalized by Aldo Leopold. But always lurking below the sandy surface are the impervious layers of lake clay and silt that retard percolation of ground water. This, plus the flat topography of the area, provides the necessary conditions for this great swamp. Cranberries, sphagnum moss, tamarack, and wildlife flourish in these wetlands, and the small amount of cultivated land is laced with drainage ditches.

SPECIAL FEATURES

Two features in the Central Plain were selected for additional description. One of these is the scenic Dells of the Wisconsin River. We also highlight the distinctive sandstone mounds and buttes that rise above the relatively flat and featureless central part of this province.

Dells of the Wisconsin River

A few miles north of Wisconsin Dells, the valley of the Wisconsin River narrows to a scenic, steep-walled canyon. For a distance of 7 miles, the Wisconsin flows between sandstone cliffs that rise nearly 100 feet above river level. Below this restriction, the river valley broadens and the channel is wider. Early French traders named this bedrock gorge "The Dalles," but with later usage it became Dells (Fig. 4-5).

Along the Dells, the cross-bedded and jointed Upper Cambrian Wonewoc Sandstone was carved by running water and etched by wind into ledged cliffs, isolated pillars and mysterious narrow chasms (Fig. 2-17). Dim and grotto-like, the mossy and moist-walled tributary canyons can vary in width from 2 feet to 100 feet, depending upon the relative resistance of the individual sandstone layers (Fig. 4-6). The general east-west orientation of the joint system controlled the trends of most of these side canyons. Potholes, attesting to the role of running water in the formation of the Dells and its tributary gorges, exist at several levels above the present elevation of the river.

The Dells of the Wisconsin River didn't exist until the Woodfordian ice withdrew from this locale. Prior to this time, the future site of the Dells was part of Glacial Lake Wisconsin, which drained to the Mississippi via the Black River. When the ice retreated, the waters of the huge glacial lake found a passage through the terminal moraine east of Wisconsin Dells. This, once again, allowed the Wisconsin to flow south and west to the Mississippi, as rapid downcutting through unconsolidated morainal material initiated the demise of Glacial Lake Wisconsin (Fig. 4-4). When the bedrock was encountered, the large volumes of escaping water cut a deep, narrow channel through the relatively soft sandstone. A Winnebago Indian legend provides a more dramatic account of the Dells' origin; it features a giant serpent whose passage across this land left the precipitous gorge we see today.

Downcutting in the area above Wisconsin Dells (Upper Dells) ceased with the construction of a power dam which raised the water level 16 feet and submerged the mouths of side canyons. The gorge below the dam (Lower Dells) is still eroding, but at a much slower rate than when Glacial Lake Wisconsin was disgorged through this narrow canyon.

The Dells are best seen from excursion boats which travel routes through both the Upper and Lower Dells daily from April through October (Fig. 4-5). The Upper Dells trip is our favorite since it includes several short stops where lithologic details may be observed more closely. These stops allow visits to Stand Rock, Witches Gulch, and Cold Water Canyon (Fig. 4-6). However, the Lower Dells are below the dam and the cliffs are higher. There are also numerous large rock islands that are carved into pleasing shapes. Especially scenic features are Hawk's Beak, Lone Rock and Sugar Bowl (Fig. 2-17). Despite crowding and overzealous commercialization, the natural splendor of the Dells of the Wisconsin is largely preserved because of ownership by the Wisconsin Alumni Research Foundation.

There are several other scenic postglacial stream channels cut into Cambrian sandstone in the Lake Delton-Wisconsin Dells area. These include Rocky Arbor State Park, which is several miles northwest of the Dells along U.S. Highway 12, and Mirror Lake State Park near the town of Lake Delton.

Figure 4-6. Witches Gulch; a deeply cut, narrow tributary to the Wisconsin River in the Upper Dells. Cross-bedding (the inclined layers) produced by currents that deposited the Upper Cambrian Wonewoc Sandstone is well shown here.

Figure 4-7. Castle Mound Roadside Park, Camp Douglas (Juneau County), Wisconsin. This intricately carved mass of vertically jointed Upper Cambrian Wonewoc Sandstone is one of the most scenic castellated buttes. Milwaukee Journal photo.

Mirror Lake and Lake Delton are the result of dams on Dell Creek. Both are beautiful, narrow water bodies surrounded by sculptured canyon walls.

Castellated Mounds and Buttes

The most obvious features of the Central Plain are the rugged sandstone mounds and buttes which rise spectacularly above the relatively flat lowland surface. Often referred to as castellated mounds because of their battlement-like profiles, they are widely scattered within the unglaciated part of the Central Plain. However, the greatest concentration occurs within the triangular area roughly defined by Wisconsin Dells, Wisconsin Rapids and Black River Falls.

Preglacial stream erosion separated these rocky remnants as outliers from the retreating Cambrian sandstone and Ordovician dolomite cuesta that lies to the south and southwest (Figs. 2-4, 4-1 and 4-2). Many mounds and buttes also suffered wave and current erosion while they were rock islands awash in the vastness of Glacial Lake Wisconsin (Fig. 2-28). The flat topped appearance of many mounds is due to a protective capping by more resistant layers in the Upper Cambrian Wonewoc Sandstone. Once this harder topping is breached, the mounds lose their square-shouldered silhouette and assume a low, rounded outline (Fig. 4-1).

Larger buttes, some of which rise more than 300 feet above the surrounding plain, include: Friendship Mound, Pilot Knob, Saddle Mound, Target Bluff, Rattlesnake Mound, Petenwell Peak, Mosquito Mound, and Sheep Pasture Bluff. Mill Bluff, a state park near Camp Douglas, is one of the units in the Ice Age National Scientific Reserve (Fig. 2-38). The mound of Roche A Cri, 25 miles north of Wisconsin Dells, is another state park (Fig. 4-3). Several other mounds are county or wayside parks, one of which is the intricately sculptured Castle Rock at Camp Douglas (Fig. 4-7).

East of the terminal moraine, a few mounds escaped the relentless march of

Woodfordian glaciation where the ice margin was thin and weak. One of these is Mt. Morris in Waushara County, but it is in marked contrast to the Driftless Area mounds. Rising nearly 250 feet above rolling ground moraine, its flanks and top are drift-strewn and modified by glacial erosion. A few sandstone ledges protrude through its glacial disguise to reveal that it once was a castellated mound. Glover Bluff (also known as Bald, Lime or Liberty Bluff) is another outlier that barely survived glaciation. This bluff, capped by Prairie du Chien Dolomite, is unusual in that it consists of faulted and folded rocks, although other mounds and the adjacent escarpment are underlain by gently dipping rocks.

Several mounds also occur along the edge of the Altonian drift in the northwestern part of the Central Plain. The rocky, steep-sided butte south of Humbird (Jackson County) is about a mile behind the terminal moraine. Glacial debris around its base suggests that this mound was a bedrock high which projected through a thin ice margin. Several sandstone ridges and craggy mounds near Neillsville in southern Clark County have a similar history, since glacial erratics are present on their lower slopes, but not on the mound tops.

Chapter 5

WESTERN UPLANDS

GENERAL CHARACTER

A northwest-southeast trending strip that includes western and southwestern Wisconsin and immediately adjacent parts of Minnesota, Iowa, and Illinois constitutes the Western Uplands Province (Fig. 3-1). Much of this region is boldly dissected by steep-sided valleys flanked by rocky bluffs that may rise as high as 500 feet (Fig. 5-1). Rising about 300 feet above the Central Plain to the northeast, the uplands slope gently to the west and southwest in response to uplift of the Wisconsin dome in Late Paleozoic time. The upland surface is composed of three major cuesta-forming units, which dip more gently than their counterparts in eastern Wisconsin. The northernmost cuesta, which underlies much of this province, is capped by the Lower Ordovician Prairie du Chien

Figure 5-1. Typical topography in the Driftless Area of the Western Uplands north of the Wisconsin River; La Crosse County, Wisconsin. The view shows a broad, deep valley carved in Upper Cambrian sandstone surrounded by dissected, retreating margins of the nearly flat upland surface of Lower Ordovician Prairie du Chien Dolomite. Wisconsin Department of Natural Resources photo.

Group (Fig. 2-4 and 4-1). South of the Wisconsin River and in northeastern Iowa, the cuesta-former is the Middle Ordovician Platteville-Galena, while the topography of the southwestern corner of this region is dominated by Silurian dolomite.

The topography provided by this series of resistant ridge-forming units is impressive since there was little Pleistocene glaciation to mask the bedrock in this province. However, there are older glacial deposits at the northern and southern ends of the province, and in southeastern Minnesota, and eastern Iowa (Fig. 2-28). The general absence of young glacial deposits and the nearly horizontal bedrock allowed major streams and tributaries sufficient time to develop an intricate dendritic network of deep valleys and flat-topped, often narrow ridges.

An exception to our previous comments on the general character of the Western Uplands is the poorly-drained, irregular, young glacial (Woodfordian) topography at the northern edge of the province (Fig. 2-28). Involved are parts of Washburn, Barron, Polk, and Burnett counties in Wisconsin, and Chisago, and northern Washington counties in Minnesota. This area is an integral part of the northwestern Wisconsin lake district that was described as a special feature in Chapter 3. The inclusion of this anomalous bit of "Northern Highland" countryside in the Western Uplands is required because the bedrock underlying the young glacial deposits in this part of the lake district is Paleozoic, and the border between these provinces is defined as the contact between the Precambrian and the Paleozoic (Fig. 2-4). This illustrates the type of difficulty encountered when one tries to divide a region into precisely bounded physical provinces. In this area, the Woodfordian ice did not discriminate between Precambrian and Paleozoic rocks, so Northern Highland topographic characteristics were firmly overprinted on the northwestern end of the Western Uplands.

A small area of Woodfordian drift projects into this province in eastern Sauk and western Columbia counties, Wisconsin. These deposits are explained in the special feature discussion of the Baraboo Range and Devils Lake. Another projection of Wisconsinan drift into the Western Uplands is shown along the southern margin of Figure 2-28. These interesting deposits are described in the glaciation section below.

Interrupting the general highland character of this province are the two great trenches cut by the Wisconsin and the Mississippi rivers. The position and elevation of these rivers sets the general pattern and depth of erosion for the entire region. Where the Mississippi, Wisconsin and their tributaries flow in softer Upper Cambrian sandstones, erosion is rapid. Such is the case throughout much of the northern two-thirds of the province, where a complicated system of narrow ridges and steep-walled valleys results (Fig. 2-4). Relatively little upland remains in this part of the province and many of the valleys are quite broad. The unglaciated southern part of this area was called "coulee country" by the early French explorers to describe the deeply cut drainages.

Where the bedrock gorge of the Mississippi River and its tributaries is within the cuestas of resistant Ordovician dolomites, as it is in the southern part of the province, stream dissection is not especially effective (Fig. 2-4). In this area, the broad southward-dipping Platteville-Galena upland is rarely interrupted by valleys. Military Ridge, extending from west of Madison into eastern Iowa, is a major topographic feature in this part of the province. This asymmetrical ridge is the drainage divide between short, steep, northward-flowing streams tributary to the Wisconsin River, and those draining gently southward to join the Rock or Mississippi rivers. Periodically, a mound capped by Silurian dolomite rises as an outlier above the surface of this upland.

The almost treeless crest of Military Ridge was utilized in 1835 as part of a military route from Fort Howard at Green Bay to Fort Crawford at Prairie du Chien. It was also an important territorial link between Milwaukee and the rapidly developing lead-zinc district. After the Civil War, a Chicago and Northwestern Railroad route was constructed along the ridge and it was said to be the longest stretch of track in Wisconsin without bridges. Later, U.S. 18 followed the same route.

The lead and zinc deposits were the lure that brought Cornish, Irish, and Italian miners from older mining areas in Missouri, and from the Old Country in the 1820s and 1830s, to found the first permanent settlements in this province. Dubuque (Iowa), Galena (Illinois), Mineral Point and Dodgeville (Wisconsin) played important roles in early mining history. In Wisconsin, the influence of mining on population distribution was so strong that the first territorial capitol was at Belmont (Lafayette County), in the heart of the mineralized area.

The Western Uplands once supported more than 2 million acres of native prairie between deciduous forests and oak openings. Clues to the locations of some of these grasslands are settlement names, such as Prairie du Sac, Prairie du Chien, La Crosse (the "Prairie" prefix was abandoned), and Muscoda (a corruption of an Indian term for "treeless land"). Now the scene is dominated by the cropland, orchards, pastures, and wood lots of attractive farms nestled in the hollows of the rolling landscape.

BEDROCK AND STRUCTURE

The paucity of Pleistocene glacial erosion and deposition in the central part of the Western Uplands provides us with an excellent outdoor laboratory to examine Cambrian through Silurian rocks, and to observe how the structure and erosional resistance of rocks influence the landscape. Rock types exposed in this region include sandstone, dolomite, limestone, and shale. All are marine deposits that accumulated as shallow seas advanced and retreated many times across this area. Cyclic fluctuations in sea level caused widespread shifts in depositional environments, which are recorded in the rocks we see today.

We previously described how this province was divisible into three general upland areas defined by cuesta forming units. The northernmost upland is developed on dolomites of the Lower Ordovician Prairie du Chien Group. This surface lies north of the Wisconsin River and continues across the Mississippi into northeastern-most Iowa and southeastern Minnesota. As such, it includes about two-thirds of the Western Uplands. This area is characterized by narrow flat-topped ridges and broader uplands associated with highly dissected lowlands carved in less resistant Upper Cambrian sandstone. The larger valleys are quite broad and relatively flat (Fig. 5-1).

South of the Wisconsin River, and extending into adjacent parts of eastern Iowa, is the Platteville-Galena cuesta which includes Military Ridge. Dipping gently southwestward, this surface is breached by streams in several places to reveal the underlying St. Peter Sandstone, and in a few instances, dolomites of the Prairie du Chien Group (Fig. 2-4).

The Silurian cuesta that cuts across northwestern Illinois into Iowa has characteristics of dip and erosional resistance that are similar to the Platteville-Galena cuesta, although it is even more resistant. Outliers from this cuesta survive as round-topped mounds well out onto the Platteville-Galena surface. They owe their existence to the durable Silurian cap rock that protects the easily eroded Maquoketa Shale beneath. Scattered examples rise above the relatively flat upland surface in Iowa, Lafayette, and Grant counties, Wisconsin, and in Jo Daviess County, Illinois. The relief of these isolated hills ranges from 200 to 400 feet. Charles Mound, just south of the Wisconsin state line, is the highest elevation in Illinois at 1,241 feet.

An oval-shaped outcropping of Precambrian rock is present at the eastern edge of the Western Uplands Province in Sauk and Columbia counties, Wisconsin (Fig. 2-4). In many ways this distinctive area is a physical province in its own right. At this locality, thick, resistant Middle Precambrian Baraboo Quartzite was folded into an asymmetrical syncline, which once was part of a more extensive foldbelt of Penokean or younger age. Erosion later isolated this elongate structure and its limbs stood as quartzite monadnocks above the general Precambrian surface. Paleozoic seas deposited sandstone and locally-derived quartzite conglomerate around and within the syncline, until ultimately the fold was buried. Late Paleozoic uplift, and the long interval that followed, allowed erosional forces to partially exhume the resistant limbs of the syncline and produce the distinctive outcrop pattern shown on Figure 2-4. The higher standing south limb forms the Baraboo Range.

GLACIATION

The most fascinating aspect of the glacial history of the Western Uplands concerns the absence, or near absence, of glacial drift in the central part of the province. Although most of the United States is unglaciated, our Driftless Area is unique in that it is encircled by glaciated country (Fig. 2-26). This region is largely confined to Wisconsin, extending only a short distance into neighboring Illinois, Iowa, and Minnesota (Fig. 2-28). The contrast between this area and the remainder of Wisconsin was first noted 150 years ago by an observer who described the absence of erratic boulders.

Although its surface was not directly modified by glacial erosion or deposition, the Driftless Area has been shaped by water and wind for a long, uninterrupted time span. The result is a region of high topographic contrast, with cliffs on steep-sided valleys, mounds, rock pinnacles, and several natural bridges. The continuous interval of weathering also left a legacy of clay-rich, cherty residual soil, solution caves, and sinkholes in areas where slightly acid water acted on carbonate bedrock (Fig. 5-2). Solution features of this type are not commonly found in the same formations in glaciated eastern Wisconsin.

There are some Pleistocene deposits within the Driftless Area but these are outwash sediments or glacial lake deposits. One extensive area of glacial lake sediments is present in south-central Wisconsin (Fig. 2-28). Although there is little evidence of deposition or erosion directly by glacial ice, it was proposed that much of the Driftless Area was invaded about 30,000 years ago by thin Altonian ice tongues. If so, this ice did little to modify the topography.

The Driftless Area is surrounded by glacial deposits of different ages; Kansan and Nebraskan are on the west, Late Wisconsinan (Woodfordian) is on the east, and older Wisconsinan (Altonian) is on the north and south (Fig. 2-28). However, the age of glacial deposits at the southeastern end of the province is a problem that is discussed in more detail below.

We observed earlier that the Western Uplands Province extends beyond the Driftless Area to include glaciated areas on the northwest and southeast. The pre-Woodfordian drift in these areas is relatively thin, deeply weathered, and nearly feature-

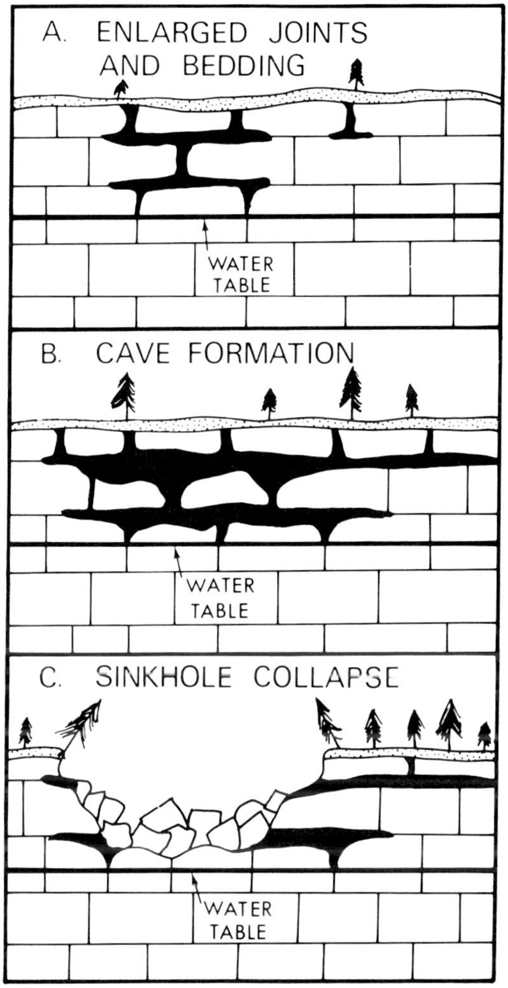

Figure 5-2. Diagrammatic cross sections illustrating the enlargement of permeable joints and bedding in carbonate rock by acidic water percolating down to the zone of permanent saturation (ground water table) to form caves and a sinkhole. In (A), joint and bedding enlargements are well developed. Caves at several levels are present in (B) and subterranean drainage may exist. The cave system has collapsed to form a sinkhole in (C). The water table in (A), (B), and (C) is shown at a constant elevation for convenience, but this is rarely the case, and ground water fluctuations will complicate cave formation.

less. The modest contrast between the earlier glaciated and nonglaciated parts of the Western Uplands can be appreciated along the Mississippi River valley by comparing the country south of Alma, Wisconsin with the area of Altonian glaciation to the north. A greater contrast exists farther north, where Altonian drift is overlain by the Woodfordian terminal moraine.

The deposits at the southeastern end of the province that are shown as Altonian on Figure 2-28 may include older tills. The glacial deposits here are wedged between the Driftless Area to the west and prominent Woodfordian end moraines to the east, and they were interpreted in several different ways by the Pleistocene experts who studied them (Fig. 5-3). As late as 1967 this entire area was considered to be Altonian, but additional study suggests that the Sugar River valley in easternmost Green County, Wisconsin may approximate the edge of the early Wisconsinan (Altonian) ice. Some scientists now consider the deposits west of the river to be Illinoian in age (Fig. 5-3). It was also suggested that Kansan deposits may be present within this area. If these interpretations are correct, glacial deposits near Brodhead and Juda, Wisconsin may contain a record of 3 major glacial stages.

Regardless of the precise age of these enigmatic, older glacial deposits, the country west of the area of Woodfordian deposition is markedly different from the recently glaciated country to the east. In the area of older glaciation, particularly west of the Rock River, the landscape is more "driftless" than glaciated in appearance. Here the drainage is well-established, glacial deposits are scarce, and the topography is influenced by differences in the bedrock.

An irregular mantle of windblown loess (silt) is draped over most of the Western Uplands. It gradually thins in a northeasterly direction, from a maximum of about 35 feet

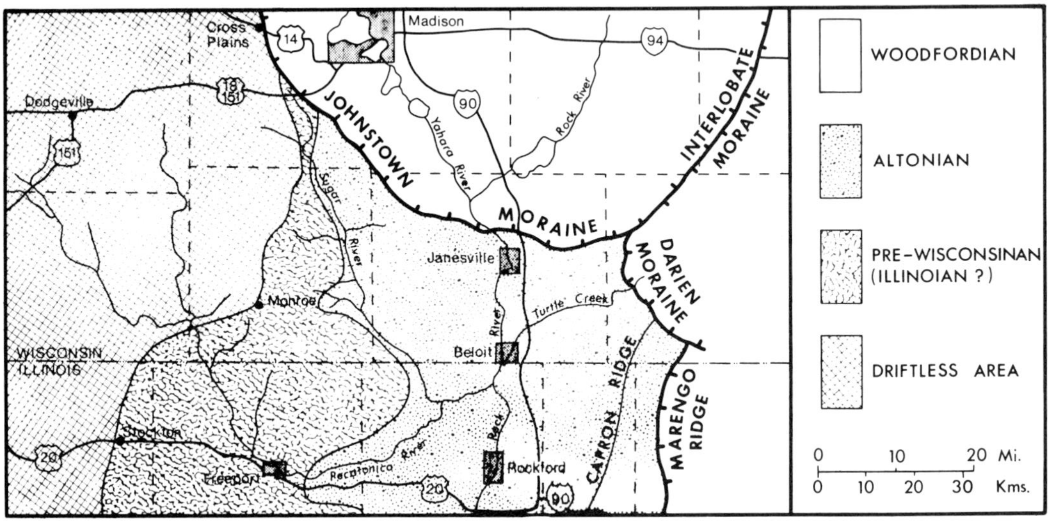

Figure 5-3. Glacial geology of a part of south-central Wisconsin and north-central Illinois. The Woodfordian (Wisconsinan) terminal moraines are labeled on the northeastern part of the map. The Driftless Area is shown on the west. In the south-central part of the map is an area that is currently interpreted to include Altonian (Early Wisconsinan) till and a pre-Wisconsinan drift that may include Illinoian, and possibly Kansan deposits. The course of the Sugar River approximates the border between these two areas of older drift.

along the Mississippi River in the southwest part of our area, to less than 2 feet along the northeastern border of the province. This airborne glacial gift greatly enriched the soil of this region.

DRAINAGE

The Western Uplands, with two major rivers and a maze of deep tributary valleys, has many physical barriers that limit transportation. The natural passages, with the notable exception of Military Ridge, are provided by the Mississippi and Wisconsin rivers, which cut across this province. Both were historic highways, and the Mississippi is still an important transportation route.

The Mississippi River is the dominant feature in the Western Uplands, and it provides some of the best scenery in the Upper Midwest (Fig. 5-4). The valley of the Great River was compared to the Rhine River gorge in Germany by David Dale Owen and Henry Schoolcraft, our pioneer geologists. To them, the rocky river bluffs suggested crumbling Teutonic castles. Schoolcraft was also impressed by a large herd of elk (wapiti) at the confluence of the St. Croix and the Mississippi back in 1821.

The valley walls of the Mississippi are steep slopes of sandstone, or rugged cliffs where more firmly-cemented sandstone or dolomite is present. The bluff height varies from about 200 to 650 feet, and the valley floor ranges from one to six miles in width. The river itself occupies only 10% to 25% of its valley bottom. The valley is widest north of Prairie du Chien, where it is cut in Cambrian sandstones. South of Prairie du Chien, the valley narrows in the resistant Ordovician dolomites. The flat flood plain contains a thin veneer of Recent clays and silts which overlie thick deposits of glacial sand and gravel. The gradient of the Mississippi is low, about 4 inches per mile, and shifting sandbars must be dredged to maintain a shipping channel. Prominent, discontinuous terraces are present along some stretches of the river, and they provide scenic sites for many villages and cities. Some terraces may be bedrock, but most are depositional features, built up and then partly cut away near the end of the Pleistocene. Natural levees also border portions of the main channel and slope gently away to the wooded backswamp regions which parallel much of the river's route.

The Mississippi River apparently developed during the Nebraskan glaciation. The position of the river valley between Nebraskan deposits and the Driftless Area suggests that the river developed at the edge of the ice during the maximum Nebraskan advance (Fig. 2-28). Prior to Nebraskan glaciation, the resistant Silurian escarpment in the southwest part of our area probably acted as a regional divide between north and south flowing rivers. Northerly drainage was blocked by the advance of Nebraskan ice, and the accumulating meltwaters drained southward across the Silurian escarpment in the vicinity of Illinois' Mississippi Palisades State Park. Here, the river cut its channel into bedrock. Once established, it maintained

Figure 5-4. Bluffs of Upper Cambrian sandstone capped by Lower Ordovician Prairie du Chien Dolomite along the Mississippi River in the vicinity of Perrot State Park, Wisconsin. The dissected nature of the flat upland surface is well shown on the Minnesota side of the river. A dolomite quarry is located at the top of the bluff on the right. Wisconsin Department of Natural Resources photo.

its general position during later glacial advances.

During the remainder of the Pleistocene, the Mississippi periodically carried tremendous amounts of glacial meltwater from the St. Croix, Chippewa, Black and Wisconsin river systems. In response to these demands, the Mississippi widened and deepened its bedrock channel. When the glaciers temporarily receded and the floods of meltwater decreased, the Mississippi backfilled and graded its valley with glacial outwash. Most of this valley fill would be removed during an interval of increased meltwater supply. When the Pleistocene ended, backfilling resulted in a great, flat valley floor with the present river lost within its broad expanse. This valley fill is 150 feet thick at Lake Pepin and 300 feet thick at Savanna, Illinois.

The Mississippi was an Indian transportation route through the Western Uplands that was followed by later explorers, trappers and traders. In 1823, the first steamboat puffed along the Upper Mississippi to open a new era. Passenger and freight traffic accelerated to a maximum in 1860, but declined rapidly after this time as competition from railroads increased. In 1925, barge-towing was initiated, and the river became an important transportation route once again. However, constant dredging is required, and a string of dams with locks is maintained by the U.S. Corps of Engineers to provide a 9-foot deep channel for barge operators. There are ten of these dams along the river within our area. In spite of these efforts, there are still navigational problems caused by shifting sandbars and variations in water level. Man does not completely control this Great River!

Numerous small streams and rivers flow into the Mississippi, but there are only a few major tributaries that pass through this province to reach the river. One of these is the Rock River, which enters the Mississippi at Moline, Illinois after draining part of the Eastern Ridges and Lowlands and the eastern part of the Western Uplands.

In our area, the longest tributary of the Great River is the Wisconsin, which flows in a large trench through the Western Uplands. Three hundred years ago, this river was part of the great canoe highway which cut diagonally southwestward across the state from the Great Lakes at Green Bay via the Fox River to the Wisconsin at Portage, and on to the Mississippi. The popularity of this route is documented in the seventeenth century journals of Raddison, Groseillers, Marquette, Joliet, and others. However, shifting sandbars in the shallow Wisconsin prohibited the development of commercial transportation.

The character of the Wisconsin River in the Western Uplands is different than it is in the Central Plain or the Northern Highland. After emerging from the terminal moraine near Prairie du Sac, it flows smoothly in a shallow channel laced with golden sandbars and tree-covered islands. The broad valley is cut in Cambrian sandstone that forms rocky walls which rise 300 to 400 feet above river level. Its gentle gradient of less than 2 feet per mile is interrupted by the power dam at Prairie du Sac, which forms the broad impoundment of Lake Wisconsin. The valley narrows just above its confluence with the Mississippi, where it is cut into the resistant Prairie du Chien (Fig. 5-5). The valley of the Wisconsin, like the Mississippi and its other major tributaries, was overdeveloped by periodic floods of glacial meltwater. Near the end of the Pleistocene, the valley floor was backfilled with sandy outwash as the flow of meltwater declined. Major tributaries of the lower Wisconsin River include Black Earth Creek, and the Baraboo, Pine and Kickapoo rivers.

Rivers which flow southward down the gentle backslope of Military Ridge include the Grant, Platte, Pecatonica, and Sugar rivers. The Grant and Platte flow directly to the Mississippi, while the Pecatonica and Sugar flow across glacial lake deposits and older drift (Altonian or older) to enter the Rock River (Fig. 5-3). The tiny Sugar River looks out of place in its large valley that was devloped by glacial torrents.

North of the Wisconsin River, the Western Upland is breached by several rivers which flow directly to the Mississippi. These include the La Crosse, Black, Trempealeau, Buffalo, Chippewa, and St. Croix rivers. The Black, Buffalo, and Chippewa were sluiceways for glacial meltwaters that carried great amounts of sediment. The East Fork of the Black was once an outlet for Glacial Lake Wisconsin. Since the Chippewa deposits more sediment than the slower flowing Mississippi can carry away, a large delta has formed. The result is a natural damming of the main river to form the broad impoundment of Lake Pepin.

Figure 5-5. Looking north across the broad, flat, sandbar-choked Wisconsin River valley from Tower Hill State Park. The flat upland surface north of the river is developed on the Lower Ordovician Prairie du Chien Dolomite. The focus at Tower Hill is an 8 foot wide vertical shaft cut 120 feet through Upper Cambrian Franconia and Wonewoc sandstones. It was used to make lead shot prior to the Civil War. Wisconsin Department of Natural Resources photo.

The Western Uplands has very few natural lakes and swamps, because of the well-developed drainage, generally porous bedrock, and the absence of young glacial deposits.[1] However, some bedrock river valleys were dammed to provide attractive water bodies. Several of these lakes are within state parks; including two in Wisconsin (Governor Dodge in Iowa County, and Yellowstone Lake in Lafayette County), and Lake-Le-Aqua-Na in Stephenson County, Illinois. Eau Galle Reservoir in St. Croix County, and Redstone and Dutch Hollow lakes in Sauk County are three other man-made lakes within this province.

SPECIAL FEATURES

In the Western Uplands there are many geologic features worthy of additional description. The ones we selected include the outdoor geologic laboratory of the Baraboo Range and Devils Lake, and the historic lead and zinc district in the Driftless Area. We also highlighted the Mississippi River by detailing the geology of several state parks situated along the Great River Road. Other featured aspects of the Driftless Area are solution caves and sinkholes, natural bridges, and the Silurian mounds.

Baraboo Range and Devils Lake

Undoubtedly, the best known and most frequently visited geologic locality in our entire region is the Baraboo Range, in eastern Sauk and western Columbia counties, Wisconsin. This area has served as an outdoor laboratory for tens of thousands of geology

1. The exception to this statement is the lakes district in the Woodfordian drift at the northern end of this province.

students and hundreds of professors over the years since the first geological observations were made here in 1852. Every spring and fall, yellow school buses clutter the roads and parking lots at the most popular sites, and outcrop study assumes some of the aspects of "take a number" and wait your turn. In spite of this drawback, it is unlikely that any geologist educated in the Midwest will fail to experience a touch of nostalgia when the Baraboo Range is mentioned, for the geology here is unusual and fascinating.

The Baraboo Range is an impressive topographic feature that rises 700 to 800 feet above the Wisconsin River, which makes a detour around it. Although the eastern half of this highland was overridden by Woodfordian ice, the western part lies in the Driftless Area. The bedrock of the range is Baraboo Quartzite, which outcrops discontinuously through a cover of Paleozoic and Pleistocene deposits. This rock is probably Middle Precambrian (Animikean) in age, although others suggest it is somewhat younger.

The general outcrop pattern of the quartzite is an east-west elongated oval, about 25 miles long and 10 miles across at its widest point (Figs. 2-4 and 4-4), which reflects the synclinal structure of the quartzite. The Baraboo Range is the moderately-dipping south limb of this asymmetrical fold. The steeply-dipping north limb generally has a more subdued topographic expression, although it is sometimes called the North Range.

The history of the Baraboo region covers a great span of geologic time, and begins with volcanism which produced the oldest rock exposed in the area. Formed about 1.8 billion years ago, this rhyolite underlies the Baraboo Quartzite. Faint flow-banding in the rhyolite establishes that it is concordant with the overlying quartzite. An unconformity probably exists between the Baraboo Quartzite and the rhyolite, but this is not proved. There are also small outcrop areas of Precambrian diorite and granite below the quartzite along the south limb of the syncline, but unfortunately the field relations of these igneous rocks cannot be established relative to either the rhyolite or the Baraboo Quartzite.

Sands that were to become the Baraboo Quartzite apparently washed off a source area to the north and were deposited in a Middle Precambrian sea. At this time the Baraboo area was only a small part of a depositional basin that extended eastward from South Dakota to at least the present shores of Lake Michigan. Eventually more than 4,000 feet of sandstone were deposited, with ripple marks and cross-bedding that attest to waves and currents in this ancient shallow ocean.

The Baraboo Quartzite grades upward into a 350 foot thick argillaceous sequence that is called the Seeley Slate. This formation, and the younger Precambrian units described below, are not exposed at the surface but our knowledge of them comes from old mining records and cuttings from well drilling. Overlying the Seeley is about 1,000 feet of the Freedom Formation; a sequence of dolomite with interbeds of slate, chert and a thick unit of hematitic iron ore. This ore was intermittently mined at three localities in the Baraboo and North Freedom areas from 1889 to 1925, but production was never significant. Two younger formations, the Dake Quartzite (200 feet thick) and the overlying Rowley Creek Slate (150 feet thick), are also part of the Precambrian succession (Fig. 5-6).

Sometime after deposition of the Baraboo and the overlying units of shale, dolomite, iron formation and sandstone, the region was subjected to intense north-south compressional forces which folded and metamorphosed these rocks. This mountain

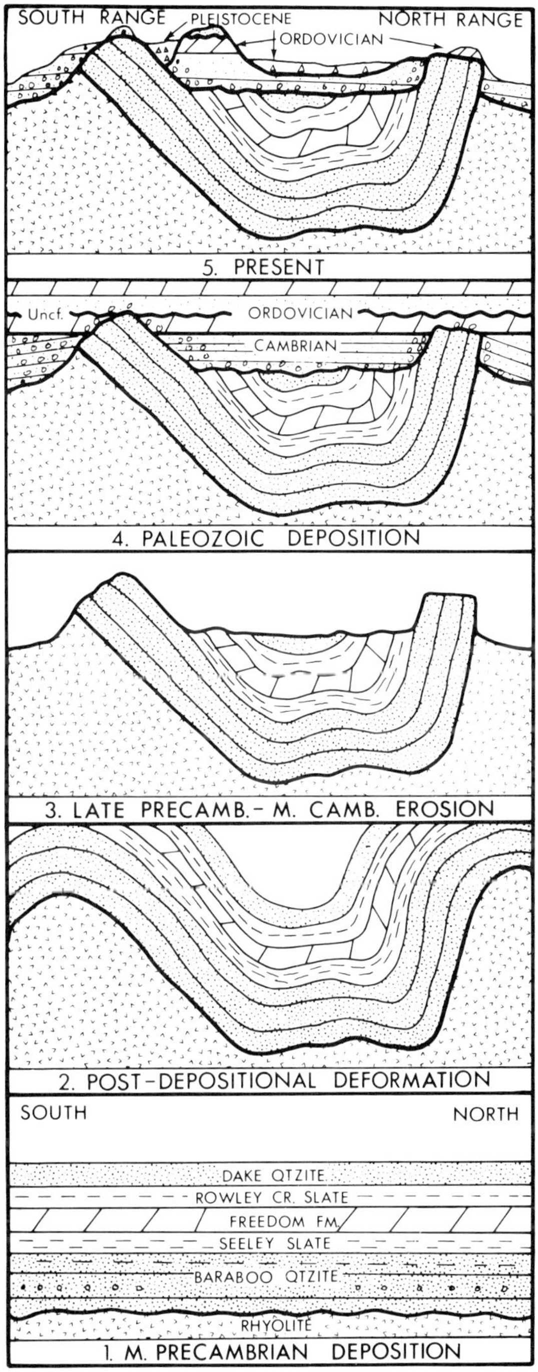

Figure 5-6. Diagrammatic cross sections summarizing the geologic history of the Baraboo syncline. The lower two panels show synclinal folding after deposition of a Middle Precambrian sedimentary succession on a basement of rhyolite. The middle panel depicts the area just prior to transgression of the Late Cambrian seas and after erosion emphasized the resistant quartzite limbs of the syncline. The panel next to the top illustrates deposition of flat-lying Paleozoic rocks around, within, and over the Precambrian syncline. Conglomeratic units are present adjacent to the quartzite, but especially south of the syncline. In the uppermost panel, erosion has removed much of the Paleozoic cover to expose the syncline as we see it today. Pleistocene glacial deposits are present in the eastern part of the syncline.

building episode may have occurred as part of the widespread Penokean orogeny that deformed Animikean rocks in the Northern Highland.

Regardless of the exact timing of this orogeny, the results in the Baraboo area were spectacular. Sandstone became quartzite and shale became phyllite[2] or slate, and the entire sequence was tightly folded into an east-west trending syncline that plunges inward from both the east and west. The major synclinal structure includes several smaller-scale anticlines and synclines at the southwest end where the outcrop expression is widest. At the narrow eastern end, the structure is a relatively simple asymmetrical fold, with northerly dips of 15 to 30 degrees on the south limb and almost vertical dips on the north limb (Fig. 5-6 and 5-7).

The presence of a well-exposed foldbelt in the Midwest has attracted many students of structural geology to this area. In addition to the major structure, the quartzite and interbedded phyllite provide excellent examples of rock cleavage and minor folds, which relate to the major structure and aid in its interpretation (Fig. 5-8).

After metamorphism and folding, the mountain system that included the Baraboo syncline was subjected to a long, intense interval of erosion that lasted until the Late Cambrian seas transgressed. During this time, the Precambrian surface was leveled to a gently undulating plain and the mountains were eroded until only isolated quartzite remnants remained. We described many of these as special features of the Northern Highland, and the Baraboo Quartzite is still another of the monadnocks that rose above the Precambrian erosional surface until buried by Paleozoic rocks (Fig. 5-6).

Figure 5-7. Ripple marks that developed on the bottom of a shallow, sandy Middle Precambrian ocean formed impressions on this bedding plane that now stands vertically on the north limb of the Baraboo syncline; Upper Narrows of the Baraboo River near Rock Springs, Wisconsin.

Late Cambrian seas transgressed around the high-standing Baraboo syncline more than 500 million years ago and slowly submerged it. At this time, it probably looked like an oval island chain in a great, shallow sea. Waves and currents, boosted by periodic storms, worked on the edges of the islands, and quartzite fragments were torn loose from the shores. These quartzite boulders, pebbles, and cobbles were added to the Cambrian sands that were accumulating around and within the atoll-like feature (Fig. 5-9). Dune deposits may also have formed along the beaches flanking the islands. Dur-

2. Most of the metamorphosed argillaceous material interbedded with the Baraboo Quartzite is not coarse enough to be called a schist. Since it lacks the well-developed cleavage of slate, it is termed a phyllite. It commonly has a silky sheen due to the secondary growth of platy minerals.

Figure 5-8. Small, asymmetrical fold in a quartzite bed within phyllite near the top of Middle Precambrian Baraboo Quartzite; West Bluff of Devils Lake State Park near the northwest entrance.

Figure 5-9. Angular unconformity between Middle Precambrian Baraboo Quartzite and Upper Cambrian boulder conglomerate and sandstone at Elephant Rock, on the East Bluff of Devils Lake State Park. The geologist is standing on the slightly irregular, eroded Precambrian surface. Boulders up to 5 feet across are visible in the lower half of the Cambrian cliff while the upper half is well-cemented sandstone.

ing this Late Cambrian transgression, sandstones of the Wonewoc and Franconia were laid down. As the waters continued to deepen, the St. Lawrence Dolomite was deposited. The sandstone formations adjacent to the Baraboo Quartzite contain spectacular sequences of boulder to pebble quartzite conglomerate that vouch for the ferocity of the stormy Cambrian sea as it attacked the quartzite islands.

These conglomeratic units are best developed south of the syncline, where they extend outward for several miles from the edges of the quartzite (Fig. 5-10). To the north, the conglomerates are much less widespread. Dalziel and Dott (1970) estimated the probable Late Cambrian position of these conglomerates relative to the equator and the inferred direction of prevailing winds. They concluded that storm waves were the effective agent in dispersing the pebbles away from their source. Their conclusion illustrates the importance of the "rare event" in geology. We tend to get bogged down in our millenia and the concept of slow, but relentless processes at work on, and within the earth. Actually, much geologic work is done instantaneously. Earthquakes, floods, and monumental storms, such as hurricanes, modify our landscape dramatically in mere moments. Apparently this was the case at Baraboo during the Late Cambrian, when occasional great storms spread rocks derived from the Precambrian highlands far and wide.

Figure 5-10. Conglomeratic units within Upper Cambrian sandstone at Parfreys Glen on the south side of the South Range. The geologist is looking at a large, well-rounded quartzite cobble. Others are visible in the layers to the right and on the opposite cliff. Wisconsin Department of Natural Resources photo.

The Late Cambrian seas briefly withdrew from this region before another transgression is recorded by the presence of the Upper Cambrian Jordan Sandstone and the Lower Ordovician Oneota Dolomite. By this time, all but the highest quartzite was covered by Cambrian sediments and none may have surfaced above the waters of the Oneota sea. The youngest Paleozoic unit in the area is represented by a few outcrops of the Middle Ordovician St. Peter Sandstone, which is unconformable on the Oneota. During the erosional interval before deposition of the St. Peter, it seems likely that some of the quartzite was exposed.

Younger Paleozoic rocks were deposited above the St. Peter, but they were eroded after a Late Paleozoic regional uplift and no record of their presence remains in this area. During the long period of Late Paleozoic to Cenozoic erosion, the Precambrian Baraboo structure was slowly exhumed to be exposed as we see it today, with essentially flat-lying Paleozoic rocks resting on steeply dipping, metamorphosed Precambrian rocks (Fig. 5-6).

The syncline is conveniently divided into

three distinctive areas. The higher and broader, relatively flat-topped, Baraboo Range (sometimes called the South Range or Baraboo Bluffs) contains Devils Lake, in the only prominent gap that cuts through this upland. The narrower, lower North Range is interrupted by 5 stream gaps, two of which are now sediment-filled and dry. Some of these gaps were eroded prior to transgression of the Late Cambrian seas, since Cambrian sandstone was deposited within them. Others may have formed later, either during the interval of erosion during the Late Paleozoic and Mesozoic, or during the Pleistocene.

Between the ranges, which loop around and connect at the ends, is an irregular lowland surface 400 to 800 feet below the surrounding ranges. This region within the syncline contains younger Precambrian dolomite, slate, quartzite and iron formation which were folded along with the Baraboo Quartzite. These formations were less resistant than the Baraboo and they were deeply eroded prior to Cambrian deposition. Cambrian and Ordovician sediments covered all surface exposures of the younger Precambrian rocks, and although later erosion has removed much of this record, they still remain concealed (Fig. 5-6).

The Baraboo Quartzite is well-exposed where it outcrops along the ranges and at the ends of the plunging syncline. It consists of medium to coarse-grained quartz that is indurated to a hard, vitreous quartzite. The quartzite is medium to thick-bedded, but in many areas bedding is indistinct and difficult to separate from well-developed joint planes (Fig. 5-11). The rock is characteristically pink to dark purple in color because of minute amounts of iron oxides. Color banding is common within the quartzite and it may be confused with bedding if one is not careful. Within the quartzite, there are numerous conglomeratic beds that contain white quartz granules and pebbles up to one inch in diameter. There are also thin interbeds of argillaceous material, now metamorphosed into phyllite. These interbeds are best developed toward the top of the quartzite where they apparently represent the beginning of a transition into the Seeley Slate. Both the conglomerate and phyllite beds provide good opportunities to determine orientation in areas where bedding in the quartzite is poorly defined.

Figure 5-11. Poorly defined beds in Middle Precambrian Baraboo Quartzite that dip north at 20 degrees are obscured by steeply south-dipping, well-developed joints on the south limb of the Baraboo syncline at the La Rue quarry near North Freedom, Wisconsin. The geologist is standing in an enlarged joint plane.

Our description of the Baraboo area lacks one final and important chapter from the past. This concerns the geologic history of the Pleistocene as recorded by the moraines and associated deposits of the Woodfordian ice, which mantle the east half of the fold. To the west of the Woodfordian terminal moraine, within and just north of the syncline, deposits of Glacial Lake Baraboo (which became an arm of Glacial Lake Wisconsin) are present. The topography of the area south of this glacial lake and west of the terminal moraine displays the classic features of the Driftless Area (Fig. 4-1). This amount of variation in Pleistocene history in one small area is another reason that Baraboo is such an outstanding outdoor geologic laboratory.

During the Pleistocene, about 13,000 to 16,000 years ago, the advancing Woodfordian ice formed a prominent end moraine which loops across Wisconsin from Hudson, on the St. Croix River, into Illinois, near Lake Geneva. In the Baraboo area, it trends southward from Wisconsin Dells and across the Baraboo syncline to Prairie du Sac (Fig. 4-4).

The Woodfordian ice moved relatively easily across the North Range and the lowland areas within, and south of the syncline. Although movement was slowed by the broader, higher South Range, ice managed to climb up and over the resistant quartzite to the vicinity of Sauk Point, about two miles east of Devils Lake. The retardation of the ice as it climbed the Baraboo Range is recorded by the terminal moraine which loops sharply around this highland (Fig. 4-4). The terminal moraine is a ridge about 20 to 60 feet in height and less than 300 feet wide.

The outer portion of the Woodfordian moraine is quite smooth, but the inner areas are knobby and pitted as is characteristic of deposits formed by stagnating ice. Where the glacier overrode quartzite, a striated and polished surface resulted. Along the north and south flanks of the Baraboo Range, the ice carried away great quantities of blocky quartzite that had piled up along the edges of the bluffs. Beyond the reach of the ice, accumulations of angular rock debris (*talus*) remain and almost conceal the bluff faces (cover photo). Aprons of sandy outwash, now used for construction and road building, were deposited by meltwaters in lowland areas bordering the terminal moraine.

The preglacial Wisconsin River flowed southward through the Lower Narrows now occupied by the Baraboo River, and across the center of the syncline to exit through the gap where Devils Lake is now located (Fig. 4-4). It initially flowed over the flat-lying Paleozoic sediments that buried the Baraboo Range. Once its channel became established in Paleozoic bedrock, the river successfully eroded downward into the underlying resistant quartzite to form two impressive gaps through the limbs of the syncline. However, there is some evidence that suggests these gaps may have formed, at least in part, during the Late Precambrian to Late Cambrian erosional interval. If so, the ancestral Wisconsin River exhumed them from their Paleozoic cover.

Prior to Woodfordian glaciation, an ancestral Baraboo River may have flowed south through the Upper Narrows at Rock Springs to join the Wisconsin within the syncline. The present Baraboo River still flows through the Upper Narrows, but now it exits northward through the Lower Narrows, formerly occupied by the ancestral Wisconsin, thus reversing the drainage direction through this gap. A few miles west of the Upper Narrows is another water gap occupied by Narrows Creek. A tributary to the preglacial Baraboo River may have once flowed south through this canyon to join the Baraboo River within the syncline (Fig. 4-4).

The advance of the Woodfordian ice disrupted the established drainage system, including the course of the Wisconsin River. When the ice retreated, the newly deposited drift and the Baraboo Range itself caused the postglacial Wisconsin River to seek an easier route southward. It looped eastward around the range to Portage, where it turned westward to return to its preglacial channel near Merrimac. As previously described, the Dells of the Wisconsin River were also developed during the postglacial rejuvenation of this river.

As meltwaters poured from the wasting Woodfordian ice front, glacial lakes formed in the Baraboo area. These lakes no longer exist, but their flat, clay, and organic-rich deposits remain. Glacial Lake Wisconsin, which flooded a considerable portion of central Wisconsin to an elevation of 1000 feet, was immediately north of the Baraboo Range. Within the lowland area of the syncline, a small glacial lake formed. It probably wasn't much more than 100 feet deep and it rose less than one-third of the way up the sides of the quartzite bluffs. This was Glacial Lake Baraboo, and it initially occupied the syncline west of the terminal moraine. As the ice retreated, the lake deepened, and it expanded eastward and northward to merge with Glacial Lake Wisconsin (Fig. 4-4). The draining of Glacial Lake Wisconsin through the Dells of the Wisconsin helped end the life of Glacial Lake Baraboo. This sudden surge of water down the Wisconsin also resulted in a short-lived Glacial Lake Merrimac south of the range.

Several small proglacial lakes formed on top of the Baraboo Range within the loop of the terminal moraine. The only evidence of these lakes now consists of flat, heavily farmed depressions with some marsh vegetation. At one time, these lakes were apparently joined together and drained into the north end of Devils Lake by way of Hells Canyon. When the Woodfordian ice retreated, they became isolated puddles as the meltwater supply ceased.

Devils Lake formed in the Baraboo Range when both ends of the preglacial gorge of the Wisconsin River were plugged by morainal debris and glacial ice. Today, till dams at the north and southeast ends of the lake impound an attractive body of water that is more than a mile long, a half-mile in width and 30 to 40 feet deep (cover photo and Fig. 5-12). This lake is the central attraction for Devils Lake State Park, one of Wisconsin's most popular, and an important unit in the Ice Age National Scientific Reserve.

At present there are no surface outlets for Devils Lake, which is fed by two small streams entering from the southwest (Messenger Creek) and northeast corners (Hells Canyon Creek). These, plus precipitation and ground water, maintain the lake surface about 500 feet below the towering East and West Bluffs, which rise vertically from steep talus slopes that reach to the lakeshore.

When first formed, this depression was filled to the brim by meltwaters, which probably flooded up the valley now occupied by Messenger Creek to cross the divide into the headwaters of northwesterly flowing Skillet Creek. When Glacial Lake Baraboo drained, Skillet Creek cut a miniature Dells at Pewits Nest, which includes several beautiful waterfalls within a narrow canyon of Cambrian sandstone.

A visit to Devils Lake is necessary to properly appreciate its event-packed geologic history. While perched on one of the bluffs, first contemplate the Precambrian orogeny, and then the Cambrian sea swirling around you. Imagine how beautiful the ancestral Wisconsin River must have been as it cut this scenic gorge, and then try to visualize an ice and moraine-dammed Devils Lake that stood about 200 feet higher than the present lake level.

With these thoughts in mind, journey along the East Bluff Trail to examine the

Figure 5-12. A northward view of Devils Lake within the south limb of the Baraboo syncline, from the East Bluff toward the beach and the Woodfordian morainal dam beyond. The well-jointed, resistant rock in the foreground is Middle Precambrian Baraboo Quartzite. Wisconsin Department of Natural Resources photo.

enigmatic potholes carved in the Baraboo Quartzite surface (Fig. 5-13). These features were cut by the swirling waters of a river, and they might predate the time when the ancestral Wisconsin first began to downcut into the Precambrian. Similar potholes a few miles away have Paleozoic sandstone plastered on their walls, so they formed prior to the Late Cambrian. It is possible that some could have developed during any one of the erosional intervals that affected this area; such as during the Late Precambrian to Late Cambrian, the pre-Middle Ordovician, the Late Paleozoic through Mesozoic, and possibly even the Pleistocene.

The East Bluff is also an excellent place to speculate on the origin of the jumbles of large quartzite blocks that make up the im-

Figure 5-13. Potholes in Middle Precambrian Baraboo Quartzite along Pothole Trail on the East Bluff of Devils Lake State Park, Wisconsin.

pressive talus accumulations that flank Devils Lake gorge and extend around the south side of East Bluff, up to the maximum position occupied by the Woodfordian ice. The angularity and unweathered appearance of the quartzite blocks, the steep slope of the talus, and the size of the accumulation argue against its existence when the Wisconsin River flowed through this gap. We also know that no significant rockfalls occurred here in historic time. It has been suggested that this talus represents a prehistoric earthquake-induced rockfall. This is an appealing mechanism but the castellated cliffs and associated monument rocks such as Balanced Rock and Devils Doorway are delicate features that never would have survived such a catastrophic event (Fig. 5-14). It seems

Figure 5-14. Devils Doorway, constructed of jointed Middle Precambrian Baraboo Quartzite, on the East Bluff of Devils Lake State Park, Wisconsin. The bedding at this locality is well defined. Wisconsin Department of Natural Resources photo.

most likely that alternate freezing and thawing during the harsh climate of the Woodfordian caused these jointed rocks to be pried loose from the steep cliffs to form these talus deposits.

Considering the well exposed and interesting geology of this scenic area, it is natural that it is a featured field trip in our companion volume.

Mounds

An assortment of bedrock mounds capped with Silurian dolomite dot the landscape of the Driftless Area of southwestern Wisconsin and northwestern Illinois. These features are stubborn left-overs from the gradual stripping of Paleozoic sedimentary strata off the Wisconsin dome, for they rest on the gently south-dipping Middle Ordovician Platteville-Galena surface, and are outliers from the southward retreating Silurian escarpment (Fig. 4-1).

The Blue Mounds of Wisconsin lie astride the Iowa-Dane county line and are the best known of these isolated bedrock remnants (Fig. 5-15). They are about 70 miles west of the Silurian escarpment in eastern Wisconsin, and 50 miles northeast of the Silurian escarpment near Galena, Illinois and Dubuque, Iowa. The western crest of these mounds is the site of Blue Mound State Park, and Brigham County Park is located on the eastern crest.

The Blue Mounds rise 300 to 500 feet above the Military Ridge divide between north-flowing tributaries of the Wisconsin River and south-flowing streams that drain to the Pecatonica and Rock rivers. When the area was first settled in about 1830, the mounds supported a typical prairie vegetation, but now they are covered with hardwoods that serve to impart the dark, bluish appearance that provides the name for this feature.

The western mound, with an elevation of

Figure 5-15. A view of the tree-covered, Silurian-capped Blue Mounds from the west. The quarry in the foreground is developed in the Middle Ordovician Galena Dolomite which forms the cuesta surface in this area.

1,716 feet, is the highest point in southern Wisconsin. It is capped by 85 feet of resistant, silicified Silurian dolomite which protects the softer Upper Ordovician Maquoketa Shale below. The lookout towers of Blue Mound State Park provide a view for miles over the Western Uplands. On a clear day, one can see the State Capitol dome in Madison, 22 miles to the east, and the Wisconsin River and the Baraboo Range to the north. The flat-topped nature of the bedrock cap proved ideal for an oval horse-racing track in the old days. The position of this track is still revealed in the pattern of the park service road at the top of this mound.

East Blue Mound only reaches an elevation of 1,490 feet and the top is a broad, flat surface developed on the Maquoketa Shale. This mound was also once topped by Silurian dolomite but little evidence of this cherty unit remains. The lack of Silurian dolomite on this mound is a bit of a puzzle. Either the dolomite to the east was locally more soluble and was removed at a more rapid rate, or some other erosional agent removed the protective dolomite cap a short time ago (geologically speaking). If the erosion of the Silurian dolomite had not been a recent event, much less of the Maquoketa Shale would remain. The Platteville-Galena bedrock surface adjacent to the mounds on the east is only half as thick as it is on the west. It was suggested that East Blue Mound and the immediate surrounding area may have been involved in an early Wisconsinan (Altonian) glaciation which removed the Silurian cap rock and eroded much of the bedrock around the mound.

Water seepage down through the Maquoketa Shale in the Blue Mounds area has produced Cave of the Mounds and Lost River Cave by dissolving part of the underlying Galena Dolomite. Both of these caves are immediately south of the Blue Mounds along U.S. Highways 18 and 151.

An assortment of similar but smaller

mounds dot southwestern Wisconsin and the adjacent part of Illinois between the Blue Mounds and the Silurian escarpment. In Wisconsin, these include Sinsinawa Mound in Grant County, and White Oak, Platte and Belmont mounds in Lafayette County. Belmont Mound is a state park and the lookout tower here provides a good view of the Driftless Area and other nearby mounds. In Illinois, examples are Scales Mound and Charles Mound. All of these mounds are capped by Silurian dolomite, but they vary somewhat in size and elevation. Lower, more subdued mounds of Maquoketa Shale are also present. These mounds, like East Blue Mound, have lost the dolomite cap rock and are now eroding away more rapidly.

Caves and Sinkholes

The solvent action of slightly acidic, percolating ground water on carbonate bedrock has produced an assortment of caves and sinkholes in the Driftless Area. It is likely that other parts of Wisconsin with limestone and dolomite bedrock also had these features, but they are either covered by Pleistocene drift, or were removed by glacial erosion.

Ground water forms caves and sinkholes by slowly enlarging bedrock joints and bedding planes as it seeps downward. As these passageways enlarge, caves form, and eventually sinkholes may result from the roof collapse of underground caverns (Fig. 5-2). There are more than 70 sinkholes reported in Vernon, Crawford, and Richland counties of Wisconsin, and many of these provide entrance to caves.

An excellent example of this type of cave development is available for examination at Maquoketa Caves State Park northwest of Maquoketa, Iowa in Jackson County. Here, several large sinkholes about 100 feet deep are connected by underground streams to form a cave system. These solution features are within the Silurian dolomite and they probably formed in the Pleistocene during periods of abundant rainfall. Commercial Iowa caves include Crystal Lake Cave (Dubuque County) and Spook Cave (Clayton County). Similar caves are present just outside our region in Filmore County, Minnesota. Examples include Niagara and Mystery caves.

Most Wisconsin caves are relatively small, due to the slow solution rate of the dolomitic bedrock. The dolomite of the Lower Ordovician Prairie du Chien Group exhibits more solution activity than the younger dolomites in the Driftless Area, but caves also occur in the Galena Dolomite. Some of the larger Wisconsin caves are commercially developed, and are available for touring during the summer months. These include: Cave of the Mounds (Dane County), Lost River Cave (Iowa County), Eagle Cave (Richland County), Kickapoo Caverns (Crawford County), and Crystal Cave (Pierce County) (Fig. 5-16).

The discovery of most caves is accidental and often steeped in legend. Since carbonate is still being dissolved by ground water, caverns may be introduced to the surface by the collapse of chamber roofs at any time. Occasionally, the sudden downward drop of a well driller's bit into a void will locate a previously unknown cavern. Cave of the Mounds was discovered during a quarry blasting operation in 1939. A bear, avoiding a hunter in 1849, claims credit for locating Eagle Cave. The Kickapoo Caverns are reported to have been used first by Indians and later as a shelter for early lead miners from the nearby diggings. Some of the earliest mining in the lead-zinc country of the Driftless Area consisted of removing ore deposits from the walls of solution caves and cavities.

Bogus Cave, in southern Richland Coun-

Figure 5-16. A myriad of stalactites hang from the ceiling of the Totem Pole Room in Cave of the Mounds, Blue Mounds, Wisconsin. The "Totem Pole" in the center of the photo is a complex of stalactites that have joined with a stalagmite on the floor of the cave to produce this column. Water on the floor and the smoothness of the lower part of the cave walls attest to the role of water in the formation of this cave. Photo courtesy of Cave of the Mounds.

ty, is named for counterfeiters who supposedly set up a printing operation in the cave about 1900. True to good cave lore, some of the money is reportedly still hidden there. Prior users of the cave apparently included Wisconsin River pirates during the 1860s, miners, and a wolverine whose remains were found 60 feet inside the entrance.

Natural Bridges

There are two significant examples of natural bridges developed in Upper Cambrian sandstones of the Driftless Area in Wisconsin. These began to form as enlargements along joints or bedding planes during the weathering process. These weaker areas were later enlarged and undercut by running water or wind. Both of these relatively fragile landforms are now protected as park sites. One is at Pier Natural Bridge County Park, 8 miles north of Richland Center (Richland County), where a vertically jointed spur of Franconia Sandstone was undercut by the Pine River on one side, and by a tributary on the other. When the sandstone was breached, the smaller stream was diverted through this new opening to the main stream. The span is 15 to 20 feet in length, and approximately 12 feet high. A local legend relates that Chief Black Hawk and remnants of his Sauk nation waded westward through this natural pass to escape the army, as they fled toward the infamous massacre at Bad Axe on the Mississippi.

A beautiful natural bridge that is 35 feet long, 30 feet high and 4 feet wide is located near Leland (Sauk County), Wisconsin at Natural Bridge State Park (Fig. 5-17). This span is cut into the Franconia Sandstone, and it is not related to any stream system. Apparently, this bridge formed by differential weathering of the bedrock by ice wedging and wind erosion. An overhanging cliff provided a cavelike opening beside this bridge that was occupied by early Wisconsinites some 13,000 years ago. The rock shelter continued to serve as a home for Indians of both the Archaic and later Woodland cultures, as indicated by sequences of fire pits, artifacts and projectile points, and bone debris from early venison roasts.

There is also a natural bridge adjacent to the Driftless Area in the area of old, thin drift at the eastern edge of the Western Uplands (Fig. 5-3). This bridge is in Krape Park at Freeport, Illinois, and it is associated with a feature called Twin Caves. These caves are in a cliff of Middle Ordovician

Figure 5-17. Our largest natural bridge is developed in the Upper Cambrian Franconia Sandstone at Natural Bridge State Park near Leland, Wisconsin. This feature is 35 feet long, 4 feet wide, and it spans an opening 30 feet high. Milwaukee Journal photo by Donald Nusbaum.

Galena Dolomite just above the bank of Yellow Creek, and they formed by scour as water swept around a bend during flood stages. The natural bridge is above Twin Caves on the cliff wall. Apparently, this feature also started out as an undercut bank that formed when Yellow Creek flowed at a higher elevation. Continuing erosion cut through to an enlarged joint plane, completing the arch. As Yellow Creek continued to erode deeper into the bedrock, the arch was left high and dry.

Great River Road and Associated Parks

The Great River Road was conceived by Congress in 1954 as a grouping of state and federal highways that closely follows the Mississippi River, from New Orleans, to its source near the Canadian border. In our area, this road provides scenic opportunities to view the Mississippi and to examine the geology of the steep, rocky bluffs that parallel the river along both sides of the valley. Several state parks along the Great River Road are ideal places to observe the changes in topography that result from variations in local bedrock. As previously described, these changes reflect the characteristics of three major cuesta-forming dolomites which outcrop along the river.

As one travels along the river from south to north through the Western Uplands, you progressively "step down" to older cuestas. At the southern edge of our area in Illinois and Iowa, massive, ridge-forming Silurian dolomite caps the river bluffs and protects the less resistant Upper Ordovician Maquoketa Shale from erosion (Fig. 2-4). These rocks, and the characteristic topography of this part of the Mississippi River valley can be viewed from Mississippi Palisades State Park. This park is about 45 miles south and east of Dubuque, Iowa along the Great River Road, which is Illinois 84 along this stretch of the river. Bellevue State Park, 24 miles southeast of Dubuque on U.S. 52 and 67, offers a similar geologic view from the west shore of the Mississippi River.

North of the Silurian escarpment, the bluff tops are protected by dolomites of the Middle Ordovician Platteville-Galena (Fig. 2-4). The characteristic rocks and topography of this cuesta are well-displayed at Wyalusing State Park, Wisconsin. This park is located just south of Prairie du Chien on a high bluff that rises 530 feet above the confluence of the Wisconsin and Mississippi rivers. Roadcuts along County Highway C (the approach road to Wyalusing from U.S. 18) expose a sequence of rocks that includes Lower Ordovician Prairie du Chien, Middle Ordovician St. Peter Sandstone, and Middle Ordovician Platteville-Galena, which is the ridge-former here. Point Lookout and Signal Point at the top of the bluff provide majestic overviews of both the Mississippi and the Wisconsin River valleys (Fig. 5-18). From here, Marquette first viewed the Great River. A network of trails within the park provides access to various rock shelters with such names as Sand and Treasure caves. There are also excellent rock exposures in Big and Little canyons and at The Knob.

Immediately across the Wisconsin River from Wyalusing, about 1.5 miles east of Prairie du Chien on Wisconsin Highway 27, is a roadcut that exposes the Lower to Middle Ordovician unconformity. At this locality, a broad U-shaped channel, cut into Willow River Dolomite at the top of the Prairie du Chien Group, is filled with St. Peter Sandstone.

Pikes Peak State Park is just across the Mississippi from Wyalusing and it can be reached by going across the river at Prairie du Chien to McGregor, Iowa, and then south on Iowa Highway 340. Roadcuts along 340 provide exposures of Ordovician rocks that include the St. Peter, Platteville, Decorah and Galena formations. The view from the top of Pikes Peak is similar to that provided at Wyalusing. Five miles north of McGregor is the historic Indian site of Effigy Mounds National Monument.

North of Wyalusing and the Wisconsin River, the resistant dolomites of the Prairie du Chien Group rise above the level of the Mississippi to become the protective cap rock for the bluffs adjacent to the river (Fig. 2-4). Below the cap rock are the relatively soft sandstones of the Upper Cambrian. Consequently, once the dolomites are breached, erosion proceeds rapidly and an intricate system of broad valleys with relatively narrow ridges develops (Fig. 2-4). In this area, the Mississippi River valley is significantly wider.

Figure 5-18. A view north from Wyalusing State Park across the Wisconsin River valley. The valley is narrower than it is in Figure 5-5, since the river here cuts into Lower Ordovician Prairie du Chien Dolomite. The rocks in the foreground are Middle Ordovician Platteville-Galena Dolomite. Wisconsin Department of Natural Resources photo.

The characteristic rock succession and topography along this part of the Mississippi River can be observed at Perrot State Park, about 20 miles north of La Crosse, Wisconsin. Here, the line of high bluffs along the Wisconsin side of the Mississippi River no longer crowds the shore. A 3.5 mile expanse of bottom land separates the main margin of the Wisconsin upland from forested outliers along the edge of the Mississippi River. These are the Trempealeau Bluffs, which were separated from the Minnesota side of the Mississippi by a Pleistocene river diversion that occurred when the original river channel east of the bluffs filled with great quantities of outwash sand and gravel as the glacial ice retreated northward. This forced a rerouting of the river through partial passages provided by streams that were western tributaries of the Mississippi before it was dammed with sediment. The diversion utilized the lower reaches of Cedar Creek and crossed a narrow divide into Big Trout Creek to return to the former channel. The Mississippi soon carved this new route into a gorge that left the Trempealeau Bluffs on the east side of the river.

Before the Mississippi changed its course, Cedar Creek flowed across the northwestern end of the Trempealeau Bluffs in the gap that now separates Trempealeau Mountain from the main mass of the bluffs. Today, Trempealeau Mountain rises in isolation about 500 feet above the waters of the Mississippi and the sloughs of Trempealeau Creek. Both the Winnebago Indians and the French fur traders referred to this landmark as "the hill that soaks in the water" (Fig. 5-19). The present name is from the last four

Figure 5-19. Trempealeau Mountain at Perrot State Park, Wisconsin was attached to the Minnesota shore until westward diversion of the Mississippi River occurred during the Pleistocene. The gap in the foreground that isolates this island from the Trempealeau Bluffs was once the mouth of Cedar Creek, a Minnesota tributary to the Mississippi. The dissected, nearly flat, upland in Minnesota is the Lower Ordovician Prairie du Chien Dolomite cuesta. Wisconsin Department of Natural Resources photo.

words of the longer French "La Montagne qui trempe a l'eau."

The Trempealeau Bluffs are now a part of Wisconsin's Perrot State Park, and are a double gift from Minnesota. The first was provided by the diversion of the Mississippi River during the Pleistocene. The other, more recent, was the donation of the park acreage to the state of Wisconsin by a resident of Winona, Minnesota, in 1918. Rock exposures in Perrot Park are accessible along the Brady's Bluff Trail. Here, the base of the sequence begins in the Upper Cambrian Wonewoc Sandstone, and extends up section to the Lower Ordovician Oneota Dolomite at the top of the bluff.

This park is also rich in history. Native occupancy is documented by several groups of Indian mounds within the park boundaries. It was also the site of a trading post established by Nicholas Perrot in 1685, and at this locality in 1689, the French formally laid claim to all the lands drained by the Mississippi River. They backed this claim up 42 years later by constructing a fort at the site of Perrot's trading post. A local Methodist minister attempted to extend the early human history of this region a bit further back by documenting that the Trempealeau valleys and bluffs exactly matched the Biblical geographical description of the Garden of Eden. A local brewer has taken this concept one step beyond, by calling the entire La Crosse area "God's Country" in a national advertising campaign.

About 60 miles upstream from Perrot, Minnesota's Frontenac State Park provides an excellent view of the enlargement of the Mississippi known as Lake Pepin. This water body is 22 miles long and 2 to 3 miles wide. It is impounded by a natural dam created by the delta of the Chippewa River, southeast of this park. As previously described, the Chippewa has a steeper gradient than the Mississippi and consequently it carries a larger sediment load. Upon reaching the slower waters of the Mississippi, this load is dropped and a large delta complex has developed. Sediments below Lake Pepin fill the original valley of the Mississippi with about 150 feet of sand, silt, and clay that accumulated as the carrying power of the Mississippi declined in postglacial time.

At Frontenac, the Mississippi River bluffs rise nearly 450 feet above the surface of Lake Pepin. On the cliffs, rocks ranging from the Upper Cambrian Franconia Sandstone into the Lower Ordovician Prairie du Chien are exposed. This park also provides an opportunity to compare the topography of this Altonian glaciated region with the nonglaciated valley of the Mississippi about 25 miles downstream.

Other state parks in the vicinity of the Great River Road include John A. Latsch in Minnesota with its three Lower Ordovician dolomite-capped bluffs of Faith, Hope, and Charity. Nelson Dewey and Villa Louis in Wisconsin are nongeologic historic sites. Merrick in Wisconsin is a riverside park noted for its beach. Wisconsin's Willow River State Park has canyon and waterfall exposures of Cambrian and Ordovician rocks.

Lead-Zinc District

The Driftless Area within the Western Uplands is the center of the Upper Mississippi Valley lead-zinc mining district. This district includes 3,000 square miles of southwestern Wisconsin south of the Wisconsin River, northwestern Illinois north of the Silurian escarpment, and a narrow strip of adjacent Iowa. Within this area are thousands of small lead mines, more than 400 zinc mines and a few small copper prospects; almost all are now abandoned (Fig. 5-20). Most of the lead and zinc deposits are within the Middle Ordovician Platteville-Galena Group, with the lead concentrated in the Galena Dolomite. The copper deposits, and minor occurrences of lead and zinc, are present in the Lower Ordovician Prairie du Chien Group at the north end of the district (Fig. 2-4).

Hot, mineralized waters circulating along paths of greatest permeability deposited sulfides of lead (galena) and zinc (sphalerite). These minerals filled small solution holes (*vugs*) and larger cavities, and replaced

Figure 5-20. An abandoned lead-zinc mine north of Linden along Wisconsin Highway 39. The tall building housed the hoist for this underground mine. Waste rock piles, such as the one on the left, still yield mineral specimens to collectors. Milwaukee Journal photo.

bedrock along bedding planes, joints and in brecciated zones. Locally, deposits seem to be controlled by a fold-related fracture pattern which gave mineralizing fluids additional freedom to move. Operators are now mining a low-grade sphalerite ore that contains some galena and minor amounts of copper sulfide (chalcopyrite).

Mining here followed a classic boom and bust pattern that related to the general economy. In times of high metal prices, or war-time shortages, mining accelerated rapidly. World War II brought an upsurge of interest in the region by large mining companies, but the area remained true to form and the mines couldn't be developed as large-scale operations. Today only a few remain in production, while others await better market conditions. Several mines also "operate" as tourist attractions during the summer.

Mining has continued sporadically in this district since 1685, to make this the oldest lead-zinc area in continuous production in the United States. The earliest miners were Indians, who removed galena for ceremonial and decorative purposes. In 1690, Nicholas Perrot was given a piece of lead ore by the Miamis, an Illinois tribe, and this focused his attention on the wealth available in this area. Perrot built a fort at East Dubuque, Illinois, and exploited Indian labor for his early mining ventures. However, by the end of the eighteenth century many enterprising Sauk and Fox Indians had their own small diggings, and smelted ore over open fires to supplement their incomes from hunting and fur trading.

Julien Dubuque secured permission from the Indians to mine in northeastern Iowa during the period from 1788 to 1810. Mining began in Wisconsin just north of Galena, Illinois in 1822. Two years later, the state's first permanent white mining camps were settled where the Indians once had lead works, at New Diggings and Hardscrabble (now Hazel Green). Galena, Illinois was chartered in 1827 and it became an important mining center with a busy waterway open to the Mississippi.

Within a few years, more white settlers came and Indian resentment grew. The population of southwestern Wisconsin expanded to 10,000, while Milwaukee was still only a trading post. During this boom, Potosi Station served as a port on the Mississippi River for the barge shipment of ore, and as a distribution point for supplies. The great lead strike in the Potosi region occurred in 1829, and eager prospectors moved up from the older mining areas near Dubuque, Iowa and Galena, Illinois.

Indian hostility in 1827, aimed primarily at the miners, resulted in restoration of the old fort at Prairie du Chien and construction of Fort Winnebago at the Fox-Wisconsin portage. In 1831 and 1832, Chief Black Hawk led his Sauks in futile attempts to stop white encroachment onto their historic Manitoumi—land of God. By 1840 the population of this area zoomed to 25,000, and the mines produced nearly 85% of the lead used in the world. The boom reached a peak around 1847, and the Military Ridge Road funneled masses of people westward, including the Paull contingent from Cornwall. Fraud and promotional schemes flourished in the district. Swindler's Ridge at Benton got its name about this time when miners tunneled through the hillside to rob the ore from mines on the other side. Obviously few heeded the words of the vanquished Black Hawk who advised: "It is a beautiful country. I loved my cornfields and the home of my people. I fought for it. It is now yours. Keep it as we did."

Ore was originally produced from shallow pits which still pockmark the area. Digging also provided homes for the early miners, and those who established residence in the hillsides were known as Badgers. This nickname is still applied to Wisconsinites. In contrast, the miners who came up the rivers into Wisconsin from Illinois in early spring and retreated southward in fall, like a somewhat despised local fish, were known as Suckers. Present-day Illinois residents prefer to be called Illini.

The first Cornish settlers moved into Wisconsin about 1832 and built limestone block houses opposite the diggings, just as they had done in the old country. Some of these houses still stand in Mineral Point along Shake Rag Street, which was named for the custom of waving a dishcloth to call the miners home for meals. The local restaurants and bakeries still feature delicious pasties, the Cornish miner's portable meat and potato pie.

The California Gold Rush, a cholera epidemic, declining lead prices and the depletion of easily accessible ores all combined to bring a rapid decline in mining activity after 1850. However, production spurted briefly during the Civil War. Soon after this war, production declined and other areas became the important lead producers. Zinc production began during the Civil War, with improvements in smelting technology. High levels of zinc production were achieved under wartime pressures during both World Wars, and modernization of facilities during World War II enabled production to remain at fairly high levels until recently.

It is an exciting district for the historian and the geologist. Consequently, we have included this area in a special southwestern Wisconsin trip described in our field guide volume. Old headframes mark the position of deep shafts into the earth, mine dumps interrupt the skyline, and early Badger holes dot the landscape. Fossil collecting is good throughout the Platteville-Decorah-Galena sequence, and in the Upper Ordovician Maquoketa Shale. Careful picking in the old mine dumps still rewards the mineral collector with the galena and sphalerite that brought the rush to this district in the 1830s and 1840s.

Chapter 6

EASTERN RIDGES AND LOWLANDS

GENERAL CHARACTER

The Eastern Ridges and Lowlands Province trends almost north-south across our area from northeastern Illinois to the Michigan shore of Lake Superior (Fig. 3-1). In northern Illinois and southern Wisconsin, its boundary with the Western Uplands cuts across southward dipping Paleozoic rocks, to follow the Woodfordian terminal moraine (Figs. 2-4, 2-28, and 3-1). From northern Dane County into central Oconto County, Wisconsin, this province is in contact with the Central Plain along the western edge of the Lower Ordovician Prairie du Chien Dolomite (Figs. 2-4 and 3-1). Through northeastern Wisconsin and across the central part of northern Michigan, the province borders the Northern Highland along the contact between Precambrian and Paleozoic rocks.

The eastern and northern borders of this province are Lakes Michigan and Superior, which add another dimension to this region. Recorded along the shorelines of these lakes is the history of their formation. More importantly, they are geologically active shorelines today, and this is reflected in a variety of scenic attractions. These include sandy beaches, bold bedrock cliffs, rocky islands, and dune fields.

As defined by these boundaries, the Eastern Ridges and Lowlands is a narrow strip of country 50 to 100 miles wide, underlain by Paleozoic rocks. In spite of profound modification by Pleistocene ice, the landscape is influenced by alternating resistant and nonresistant Paleozoic sedimentary rock units which essentially parallel the Lake Michigan shoreline (Fig. 2-4). The more durable units form two cuestas, flanked by lowlands developed on less resistant bedrock. The Lower Ordovician Prairie du Chien and the Silurian dolomite are cuesta formers, while the Upper Cambrian sandstones, the Middle and Upper Ordovician units (St. Peter Sandstone, Platteville-Galena Dolomite and Maquoketa Shale), and the Devonian formations occupy lowlands (Figs. 2-4 and 6-1).

The cuestas have relatively steep, west-facing scarps and long, gentle backslopes that dip eastward off the Wisconsin arch into the Michigan basin (Fig. 6-1). In upper Michigan, these cuestas trend to the east and dip to the south as they swing around the north end of the Michigan basin. In southern Wisconsin, the trend of the Paleozoic rocks curves to the west and the dips are to the south as they cross the Wisconsin arch (Fig. 2-4).

Although the bedrock plays a role in the character of this province, it is not dominant here as it was in the Western Uplands. The Late Wisconsinan ice (Woodfordian and Valderan) planed off the bedrock highs and filled in the lows to create a youthful glacial landscape. Glaciation simplified the cuestas by eliminating outliers and irregularities along their fronts, and by plastering them with drift. Rock exposures are present only where the drift is thin. Glacial scour of the bedrock was generally deep, and this may ex-

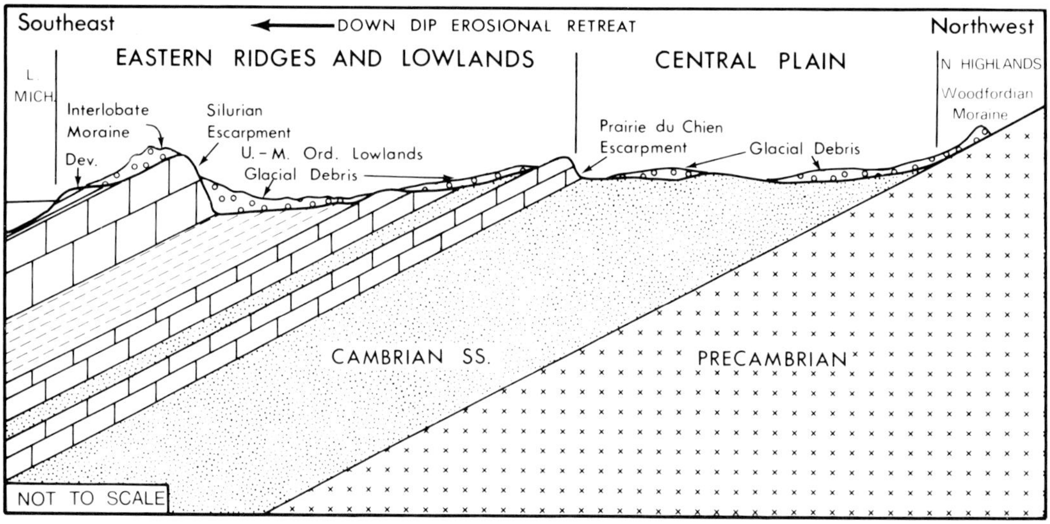

Figure 6-1. Diagrammatic cross section of the Eastern Ridges and Lowlands and the Central Plain illustrating the formation of cuestas and lowlands by the downdip erosion of strata that dip more steeply than those illustrated in Figure 4-1. Another contrast with Figure 4-1 is provided by glacial erosion and deposition. In this illustration, mounds are absent in both the Eastern Ridges and Lowlands and the Central Plain and cuestas are relatively subdued. Lake Michigan is on the far left.

plain why so few caves are present. Richardson Cave, just within the Woodfordian terminal moraine in Dane County, and Maribel Caves in Manitowoc County, are exceptions.

The glacial deposits of the Eastern Ridges and Lowlands are gently rolling to modestly hilly, unless moraines or drumlins are encountered. Many of the moraines are characterized by highly irregular topography, and local relief may exceed 150 feet. This is especially true of the Woodfordian terminal moraine from south of the Baraboo Range into northern Illinois. Wisconsin's Kettle Interlobate Moraine, trending south-southwest from northern Manitowoc County to Walworth County, is the premier glacial landform of the Eastern Ridges and Lowlands (Figs. 2-28, 6-1, and 6-2). Clusters of drumlins dominate the landscape in parts of Menominee County, Michigan and in the Wisconsin counties of Dodge, Jefferson and Dane. There are also some areas of relatively flat ground underlain by glacial lake deposits or outwash.

The drainage of this entire province is in the "young upstart" category, compared to the efficient, older systems of the Driftless Area. Glaciation dotted southeastern Wisconsin and northeastern Illinois with kettle lakes where terminal moraines and pitted outwash were deposited, and preglacial streams were destroyed, dammed or diverted (Fig. 2-35). Numerous lakes, swamps and large areas of muck soil reflect this disruption. These areas of poor drainage provided transportation problems in the days of early settlement, and hundreds of miles of toll roads made of wooden planks were constructed to provide firmer footing.

The development of fertile soils on glacial deposits in many parts of eastern Wisconsin and northeastern Illinois provided a good agricultural region adjacent to the modifying climatic influence of Lake Michigan. Early immigrants, who flocked here to farm, enriched the culture and founded communities such as Luxemburg, Belgium, Poland, Wales, Genoa City, Lake Geneva, and Germantown. In spite of their efforts to clear and till the land, morainal areas and the steeper parts of drumlins still retain deciduous forest cover.

Most of the population of our region lives and works within this province. However, the distribution isn't uniform, and there are tracts of state and national forest, large areas with family farms, and small settlements interspersed with a few prosperous cities. The large, industrialized cities cluster along Lake Michigan, the northern Fox River valley and the middle reaches of the Rock River.

The Great Lakes add a maritime flair to the agricultural and industrial Midwest. Commercial and charter fishing, busy ports linked to the seas of the world, and ship building are all part of this scene.

BEDROCK AND STRUCTURE

The lowland and cuesta pattern of this province was previously described (Fig. 6-1). The oldest cuesta-former is the Lower Ordo-

Figure 6-2. Holy Hill Monastery of the Carmelite Fathers occupies a large kame in Washington County, Wisconsin. The public observation platform in the south steeple of this church provides an overlook of the wooded, rolling kame and kettle topography of the Kettle Interlobate Moraine. Milwaukee Journal photo by Allan Y. Scott.

vician Prairie du Chien Dolomite, which forms a narrow, subdued, and discontinuous ridge. It rises slightly above the less resistant Upper Cambrian sandstones along the western margin of the province. In northern Michigan, an equivalent unit (Au Train Formation) swings eastward to form an abrupt escarpment along the Lake Superior shoreline. Part of this cliff is the scenic Pictured Rocks National Lakeshore, east of Munising in Alger County. In southern Wisconsin, the Prairie du Chien cuesta turns westward and broadens as it rises onto the Wisconsin arch (Fig. 2-4). To the east of this cuesta, a wide region is underlain by less resistant Middle to Upper Ordovician sandstone, dolomite and shale. These rocks form a relative lowland that extends from the Upper Peninsula, under Green Bay and south through areas occupied by Lake Winnebago, Horicon Marsh and the Rock River valley.

The Platteville-Galena dolomites in this province are not cuesta formers as they are in the Western Uplands. A number of factors combine to produce dissimilar roles for the same rock unit. In the Western Uplands, the dip of the Paleozoic rocks is at a lower angle so the outcrop width is greater (Figs. 4-1 and 6-1). Also, in the Driftless Area, a thick section of poorly resistant, Middle Ordovician St. Peter Sandstone is the slope former between the Middle Ordovician Platteville-Galena and the Lower Ordovician Prairie du Chien cuestas. The St. Peter in the Eastern Ridges and Lowlands is too thin, or is not present, to provide the weathering contrast necessary for a distinctive cuesta. Another important aspect is that glacial erosion reduced much of the bedrock irregularity within this province.

Overlying the Platteville-Galena in this lowland area is the Upper Ordovician Maquoketa Shale. The greenish shale, with its thin interbeds of fossiliferous limestone, occurs in a narrow belt along the base of the resistant Silurian dolomite (Fig. 2-21). However, the outcrop belt widens in northern Michigan where the entire Stonington Peninsula (across Little Bay de Noc from Escanaba) is composed of an equivalent formation called the Richmond. Exposures in eastern Wisconsin and northeastern Illinois are sparse, since the formation is not resistant. At the old Stockbridge clay pit east of Lake Winnebago, it was so deeply weathered that horse-drawn scrapers were used to mine it for brick clay at the turn of the century.

The Neda Formation, a thin, oolitic hematite, is locally present as lenses between the Maquoketa and the overlying Silurian dolomites. It only outcrops in Brown and Dodge counties, Wisconsin, and just southwest of the Chicago area along the Kankakee River. The Neda was mined as an iron ore in the vicinity of Mayville (Dodge County), Wisconsin from about 1848 to 1927 (Fig. 2-22).

East of the Ordovician lowland is the broad Silurian dolomite cuesta, which is part of the Niagaran escarpment described in Chapter 2. It forms the rugged headland of the Garden Peninsula in Michigan and continues southwestward as a string of islands in Lake Michigan until it rises as the rocky spine of the Door Peninsula in Wisconsin (Fig. 2-23). From here, it trends south as a well-defined feature along the east side of Lake Winnebago (where part of it is High Cliff State Park), and Horicon Marsh. It persists southward less boldly until it disappears under glacial drift north of Milwaukee. Occasional exposures establish its presence beneath the glacial drift in a broad band that parallels Lake Michigan as it extends into northeastern Illinois.

The rocks of the Niagaran escarpment are light-gray to brownish-gray dolomites of Early and Middle Silurian age. An interesting aspect of the Silurian in this province is the presence of coralline reefs which developed locally in the shallow, warm Silurian sea. These reefs are discussed in more detail as a special feature of this province.

Along the Lake Michigan shore between Sheboygan and Milwaukee, is a narrow lowland underlain by Middle Devonian dolomite and shale (Figs. 2-4 and 6-1). These are the youngest rocks in our area and they vanish eastward under the surface of Lake Michigan, only to emerge on other flanks of the Michigan basin (Fig. 2-3). The Devonian formations recognized in Wisconsin are, from oldest to youngest, Lake Church, Thiensville and Milwaukee. The Lake Church and Thiensville are dark-gray to brown dolomites that outcrop in a few places in Sheboygan, Washington, Ozaukee, and Milwaukee counties. The Lake Church can be seen at the quarry lake in Harrington Beach State Park north of Port Washington.

The Milwaukee Formation is a nonresistant, fossiliferous, gray-brown to buff dolomite with interbedded gray shale and limestone. It outcrops along the Milwaukee River at Estabrook Park and in a few places along the Lake Michigan shoreline (Fig. 2-24). At these localities, fractures and vugs contain natural asphalt; Wisconsin's most extensive petroleum deposit to date. This formation was quarried in northern Milwaukee County as a natural cement rock around the turn of the century.

Rock excavated from the floor of Lake Michigan during construction projects immediately offshore from Milwaukee provides the only information on the Upper Devonian Kenwood Formation. This dark shale contains a variety of fossils, including fish remains and plants in thin, coaly seams.

There are also a few small exposures of Precambrian rocks within the Eastern Ridges and Lowlands. The largest of these consists of ledges of Animikean Waterloo Quartzite in southeastern Dodge and northwestern Jefferson counties, Wisconsin. Several small pegmatite dikes cut the eastward-dipping quartzite at this locality. Otherwise it is similar to the Baraboo Quartzite which outcrops 35 miles to the northwest in the Devils Lake area. Other small Precambrian exposures in this province include several knobs of rhyolite that protrude through the Paleozoic and drift cover of southeastern Green Lake County, Wisconsin.

GLACIATION

Nowhere in the United States has Pleistocene continental glaciation created such a remarkable landscape as it has in the Eastern Ridges and Lowlands. The Woodfordian Substage, from about 22,000 to 12,500 years ago, represented a major advance of the Wisconsinan Green Bay and Lake Michigan lobes. The Lake Michigan lobe moved southward down the present Lake Michigan basin more rapidly than the Green Bay lobe advanced southwestward along the Middle-Upper Ordovician lowland. Apparently the resistant Silurian escarpment, the spreading Lake Michigan lobe on the east and the Northern Highland on the west restricted the Green Bay ice to a relatively narrow tongue. During this advance, Green Bay was not eroded as deeply as the neighboring Lake Michigan basin, but the Platteville-Galena surface was planed off to develop basins for Lake Winnebago and Horicon Marsh.

As the Woodfordian ice advanced, the resistant cuestas were reduced in size and most irregularities were removed. The ice polished and striated the gentle back-slopes of these cuestas, and this surface is visible almost everywhere in eastern Wisconsin when the drift is stripped back to expose fresh bedrock.

Prominent Woodfordian terminal moraines mark the southwestern margin of this province. These moraines are easy to recognize since they are adjacent to the Driftless Area or to areas with older, more weathered drift (Figs. 2-28 and 5-3). Such a section of terminal moraine, situated 10 miles west of Madison near Cross Plains, is part of the Ice Age National Scientific Reserve (Fig. 2-36). Here the terminal moraine rests on weathered Prairie du Chien Dolomite and overlooks the Driftless Area of the Western Uplands. Glacial erosion was not especially effective in this area since the ice was moving at an angle to the east-west strike of the bedrock. This condition also slowed the advance of the ice tongue in this area.

The Eastern Ridges and Lowlands was largely covered by Woodfordian ice, and it is generously blanketed with drift. Most of this drift is arranged in recessional or terminal ridges, formed into a prominent interlobate moraine, distributed randomly as ground moraine, or shaped into drumlins. Some relatively flat areas are veneered by outwash sands and gravels or glacial lake deposits. Several extensive areas of glacial lake deposits within this province are described below.

Broad expanses blanketed with fine outwash sediments are present in parts of this province. One of these is in southern Wisconsin and the adjacent part of Illinois, to the west of the Woodfordian terminal moraine (Fig. 2-28). Located mainly in southeastern Rock County, Wisconsin and northwestern DeKalb County, Illinois, it extends westward into the Rock River valley. This Late Wisconsinan outwash was deposited over Altonian till that continues into the Western Uplands (Fig. 5-3). A smaller outwash plain is present south of Oconomowoc, Wisconsin, along Interstate 94, and there are others scattered through eastern Wisconsin. Streams carrying great quantities of meltwater and sediment flowed southwest from the Green Bay and Lake Michigan lobes to veneer the preexisting landscape in many places with up to 300 feet of water-washed debris.

There are several large swarms of streamlined drift hills or drumlins in the region invaded by the Green Bay lobe (Fig. 2-30). Eighty per cent of Wisconsin's drumlins are found in the depositional area of the Green Bay ice, and the drumlin fields of southeastern Wisconsin, particularly those of Jefferson County, are some of the best examples. Another well-developed drumlin field is in Fond du Lac County near Campbellsport, Wisconsin, and part of it is included in the Ice Age National Scientific Reserve (Fig. 2-36).

The blunt "noses" and the long axes of drumlins reveal the direction of movement of glacial ice (Figs. 2-28 and 2-30). Flow patterns determined from drumlin orientations are especially clear for the Green Bay lobe of the Woodfordian ice in southeastern Wisconsin and for the Valderan ice in Menominee County, Michigan (Fig. 2-28). The distribution in these areas suggests that drumlins may develop in a zone of moderate ice thickness several miles behind the front, but not under the area of greatest thickness. The composition of drumlins is variable and it indicates that they represent the sculpturing of earlier deposited till and outwash during a glacial readvance.

We previously noted that the dominant glacial feature of eastern Wisconsin is the south-southwestern trending Kettle Interlobate Moraine, a depositional complex that rises 100 to 250 feet above adjacent ground moraine and outwash lowlands (Figs. 2-28 and 6-2). This conspicuous feature initially formed during the Woodfordian Substage from merged terminal moraine deposits of the Green Bay and Lake Michigan lobes as they advanced southward along parallel routes. When the Woodfordian advance stalled and the two lobes began to wane, the widening gap between them became a trough which received great quantities of meltwater and sediment from the thinning ice fronts. Outstanding examples of stagnant ice features such as kames, eskers, crevasse fills and kettles were developed during this time (Figs. 2-33 and 2-34). These are especially well-displayed in the Northern Kettle Moraine unit of the Ice Age National Scientific Reserve (Fig. 2-36). Hence, the geographic name for this morainal complex reflects its development between adjacent lobes while it also honors the magnificent display of kettle and kame topography.

Most of the glacial drift that makes up the Kettle Interlobate Moraine is locally derived. Angular and rounded Silurian dolomite cobbles and pebbles are prevalent, but a variety of Precambrian igneous and metamorphic rocks from northern locales are also present.

As the Woodfordian ice retreated northward, a series of distinctive recessional moraines formed in northeastern Illinois and southeastern Wisconsin. Since these younger moraines are essentially parallel to the Lake Michigan shoreline, they are known as Lake Border Moraines. Retreat continued and Glacial Lake Chicago formed and filled with meltwater (Fig. 2-40). Evidence of this high lake stand is recorded by silts and sands deposited over till in the vicinity of Racine, Wisconsin and Chicago, Illinois (Fig. 2-28). As the ice continued to retreat, new outlets opened for ancestral Lake Michigan and its level dropped. During this time, a boreal forest developed on the recently exposed lake deposits and spread northward with the retreat of the ice margin.

A rise in the level of ancestral Lake Michigan inundated nearshore lowlands once again, and the interglacial forest was flooded and partially buried under lake sands. The remains of a part of this forest, suggesting a climate not too different from today's, are well-preserved in a bluff along Lake Michigan near Two Creeks, Wisconsin, in northernmost Manitowoc County. This is the type locality for the Two Creekan Substage and it is also part of the Ice Age National Scientific Reserve (Fig. 2-40). A layer of reddish till overlies the lake deposits here, and records the Valderan ice advance. Any trees still standing were snapped off or uprooted by this ice tongue as it advanced to deposit up to 12 feet of till. Carbon-14 dating of spruce logs places the destruction of this forest at 11,850 years before the present.

The Two Creeks buried forest is of special interest because it provides an accessible example of a glacial advance, retreat, and readvance. A nearby quarry at Valders, Wisconsin also provides evidence of multiple glaciation. In the Valders Lime and Stone Quarry, there are two different tills, and the glacially polished Silurian bedrock records two distinct sets of striae. The older, Woodfordian till, is gray and gravelly, and it is associated with deep striae that trend essentially straight south (175 degrees) (Fig. 2-39). The younger, red Valderan clay till relates to a less well-developed set of striae that are almost at right angles (260 degrees) to the older set.

The Valderan ice also advanced over the northern end of the Woodfordian Kettle Interlobate Moraine. The glacier deposited a thin layer of drift over this prominent feature from central Manitowoc County northward, and consequently the northern end is more subdued in character than it is to the south.

The Valderan was the shortest and last glacial invasion of the Wisconsinan Stage. During its maximum advance, a relatively thin Valderan ice tongue moved down the Lake Michigan basin as far as Milwaukee, while another thin lobe reoccupied the upper part of the Green Bay lowland. As a result, drainage was again restricted by the Lake Michigan lobe and a new stage of Glacial Lake Chicago formed (Fig. 2-40). This lake, like the original Lake Chicago, drained southwest via the Des Plaines and Illinois rivers to the Mississippi.

Glacial lake sediments, extending from the present shores of Green Bay southward to Lake Winnebago and into adjacent areas on the northwest, were deposited by meltwater impounded behind the retreating ice margin of the Valderan Green Bay lobe and the irregular topography of the Valderan terminal moraine (Fig. 2-28). This impoundment was Glacial Lake Oshkosh, although a precursor had a Woodfordian history that we related in Chapter 4 on the Central Plain. During the Valderan ice retreat, Lake Oshkosh drained southeastward to an ancestral Lake Michigan over low places along the Silurian escarpment. As the Valderan ice continued its northeastward withdrawal, Glacial Lake Oshkosh expanded, and more northerly outlets were utilized. These outlets were the valleys now occupied by the Manitowoc, Kewaunee and Anhapee rivers. The gap at Sturgeon Bay that separates the upper part of the Door County Peninsula from the mainland was also a drainageway late in the history of this lake. Eventually Green Bay became ice-free, and Lake Oshkosh spilled northward by way of the Fox River. With time, most of Lake Oshkosh was drained, although irregularities on its floor contain several water bodies, including Lake Winnebago, which still drains down the Fox to Green Bay. At its maximum level, Glacial Lake Oshkosh stood 65 feet above present-day Lake Winnebago.

Late in the Valderan, the Lake Superior basin was largely open, although ice remained in the northeastern part of the Lake Michigan basin and in the eastern part of the Upper Peninsula of Michigan. This forced ancestral Lake Superior to drain southward across the center of the Upper Peninsula by channels that included the one now occupied by the north-flowing Au Train River and the south-flowing Whitefish River. Au Train Bay on Lake Superior and Little Bay de Noc near Escanaba, Michigan were also parts of this glacial spillway (Fig. 2-40).

DRAINAGE

The drainage systems of the Eastern Ridges and Lowlands are youthful and inefficient compared to those of the Western Uplands and most of the Central Plain. Although drainage patterns in this province are influenced somewhat by bedrock, they are mainly controlled by Late Wisconsinan glacial erosion and deposition.

The river systems of this province with Atlantic-bound waters include: streams in the Upper Peninsula that drain north to Lake Superior, the longer south-flowing streams of the Upper Peninsula that enter Lake Michigan, the rivers that cross this province to enter Green Bay or Little Bay de Noc, the northern Fox River system that goes northeast to Green Bay, and the short streams that flow eastward into Lake Michigan. River systems whose waters reach the Gulf of Mexico via the Mississippi include: the Rock River system, the southern Fox River, and the man-made drainage of the Illinois waterway.

The streams of northern Michigan that enter Lake Superior within this province are all short, since the drainage divide is controlled by the resistant, south-dipping cuesta of the Lower Ordovician Au Train Formation, which closely parallels the Lake Superior shoreline (Fig. 2-4). Waterfalls occur where streams drop over this escarpment. These falls include Laughing Whitefish, Au Train, Munising, and Miner's, to name a few (Fig. 2-19). Drainage southward to Lake Michigan flows downdip off this cuesta and crosses younger Ordovician rocks. Such rivers include the Sturgeon and Whitefish. The Whitefish River served as an outlet for ancestral Lake Superior until the present-day outlet at Sault Ste. Marie became ice-free.

The larger rivers that flow easterly across this province to enter Green Bay or Little Bay de Noc rise in the Northern Highland of Wisconsin and northern Michigan. These include the Escanaba, Ford, Menominee, Peshtigo, and Oconto rivers. Before Pleistocene glaciers formed Green Bay and Lake Michigan, the Menominee, Peshtigo, and Oconto may have merged in the area now occupied by Green Bay. Then they could have drained east to carve a passage in the Silurian escarpment at Sturgeon Bay on the Door Peninsula. If so, the waters of Glacial Lake Oshkosh that flowed through this same cut during the Valderan retreat merely adopted this older channel. This adoption procedure was continued in 1878, when this gap was blasted below lake level to eliminate the narrow portage formerly required. A later enlargement resulted in the present Sturgeon Bay Ship Canal.

The upper Fox River rises near Portage in the Central Plain and flows northeastward to Lake Winnebago within the Eastern Ridges and Lowlands. All of the major tributaries to the Fox rise in the Northern Highland and flow across the Central Plain. These tributaries and the upper reaches of the Fox were previously described.

Lake Winnebago is the largest inland lake in Wisconsin. It is a remnant of Glacial Lake Oshkosh which had a long and complex history, as previously described. A late stage of Glacial Lake Oshkosh was impounded behind a dam of glacial drift. The lower Fox River, downstream from Lake Winnebago, easily cut through this till barrier into the underlying bedrock to create a drop of 15 feet per mile from Neenah-Menasha to Kaukauna. Eight rapids marked the course of the Fox between Neenah-Menasha and De-Pere, and these influenced settlement. Pere Claude Allouez established a mission on the Fox at the first rapids upstream from Green Bay in 1671. These rapids were called "Rapides des Peres," which later was shortened to DePere. Manufacturing centers developed at the various rapids, because they were good hydropower sites. The first use of running water to generate electricity in the United States was on the Fox River at Appleton in the early 1880s.

Green Bay, at the mouth of the Fox, was the first permanent settlement in Wisconsin. Jean Nicolet landed along the shore at Red Banks about 10 miles northeast of here to claim this region for France in 1634. Welcomed by the Indians, he reported that the main course of a feast given in his honor consisted of 120 beaver. Fort La Baye was constructed at what is now Green Bay 49 years later, and surprisingly enough, beaver were still abundant.

The Fox River has suffered through 300 years of use and abuse by the white man. Marquette and Joliet explored the entire length of the Fox in 1673 before portaging to the Wisconsin and traveling on to the Mississippi. Once the area's busiest and most exciting transportation route, its chain of locks is still maintained for small boats. Hydroelectric dams and associated industrial developments, primarily paper mills, turned the free flowing Fox River of Nicolet, Marquette and Joliet into one of the more polluted rivers in the nation. Today, serious efforts are being made to clean up the Fox before it destroys all of Green Bay.

Streams in Wisconsin and northern Illinois that drain into Lake Michigan flow downslope off the divide provided by the Silurian cuesta. These relatively short rivers include, from north to south, the Ahnapee, Kewaunee, Twin, Manitowoc, Sheboygan, Milwaukee, Root, Pike, and Chicago rivers. Of these, the Manitowoc is the only one which cuts completely across the Silurian cuesta. This breach in the escarpment probably developed when overflow through a notch helped to drain an early stage of Glacial Lake Oshkosh.

The courses of many of these rivers are indirect, since they are strongly influenced by irregularities of the Pleistocene surface. This is especially true near Lake Michigan, where some tend to flow parallel to the lakeshore because recessional moraines border the lake. The Milwaukee River, for example, closely parallels Lake Michigan for the last 32 miles of its length, before reaching an opening in the barrier moraines.

There are several large marshes on the backslope of the Silurian cuesta in Wisconsin, that once were glacial lakes which gradually drained and filled with vegetation. Some of these marshes form the headwaters of the rivers previously listed. Sheboygan Marsh, one of the units in the Ice Age Scientific Reserve, was a lake covering 16 square miles to a depth of 45 feet. Elkhart Lake is a remnant of this water body. A similar feature, east of Chilton on the Manitowoc River drainage, is another extinct glacial lake.

The Kettle Interlobate Moraine and the Silurian dolomite cuesta combine as the drainage divide separating the southern Fox River system from the Rock River to the west. The Fox has its headwaters just east of the Kettle Moraine lakes at Oconomowoc, Wisconsin. The Pewaukee Lake basin is part of this system, as are lakes Delavan, Geneva, and Beulah, to name some of the larger ones. Confined by the Lake Border Moraines, the southern Fox flows through the southeast Wisconsin lakes region, enters Fox Lake in Illinois, and then continues south to the Illinois River. A similar southward path is followed by the Des Plaines River, which heads in Wisconsin and parallels the morainal systems to the Illinois.

The Rock River is another drainage in this province that contributes its water to the Mississippi, after crossing into the Western Uplands. The lower reaches of this river, and its tributaries which drain the Western Uplands, were previously described. The upper Rock and its major tributaries (the Yahara, Crawfish, and Bark rivers) flow across the recently glaciated Eastern Ridges and Lowlands. They are all low-gradient streams that drain marshy areas dotted with lakes. The Yahara, for example, originates in northern Dane County and acts as the interconnecting drainage between the four lakes in the Madison area.

The headwaters of the Rock are in Fond du Lac County, southwest of Lake Winnebago. It flows along the Ordovician lowland through Horicon Marsh, Lake Sinissippi and Lake Koshkonong. Deviating westward from its preglacial course, it flows past Janesville to Beloit. From Beloit to about Dixon, Illinois, the Rock follows the eastern border of the Western Uplands until it cuts westward to the Mississippi.

Horicon Marsh occupies a shallow glacial basin that was once a respectable, but shallow, lake three times larger than present-day Lake Mendota at Madison. It gradually filled with sediment and vegetation until a dam was built on the Rock in 1842 to create a higher water level. Ever since this time, human manipulation of Horicon has continued. A court decision in 1914 allowed ditching and drainage of the area despite the violent protests of hunting clubs, which enjoyed excellent shooting on the marshy lake. Waterfowl habitat was destroyed when the area was drained, and farming proved unsuccessful on the poorly drained, peaty soils. The bankrupt farms came under public ownership in the early 1940s and the water level was raised once again to restore wildlife habitat. More recently the area was temporarily drained in an attempt to poison the rough fish. State and federal wildlife refuges on the marsh attract ducks and hundreds of thousands of migratory geese. This excessive concentration of geese is the current problem at Horicon.

During the early rush of immigration and settlement in Wisconsin, a plan was devised to connect the Rock River near Jefferson, via Oconomowoc Lake, to Lake Michigan at Milwaukee. This route would require a multitude of locks to manage the several hundred feet of elevation difference between Lake Michigan and Pewaukee. More than 200,000 acres of federally-owned land was granted to the Wisconsin Territory to help finance the canal, which had an estimated cost of one million dollars in 1836. Some work actually began before this ill-conceived project was abandoned.

In the Chicago area, a man-made cut across the Silurian drainage divide diverts some Lake Michigan water from its North Atlantic heritage to the Gulf of Mexico. This cut is the Chicago Sanitary and Ship Canal and it connects the West Branch of the Chicago River with the Des Plaines River, which flows to the Illinois. This route, from the end of the Silurian cut to the Mississippi, is the overflow channel utilized by high stands of Glacial Lake Chicago. The rejuvenation of this ancient drainage across Illinois was opened in 1933 as the Illinois Waterway, and it is now one of the busiest inland water systems in the world. A 9 foot channel from Lake Michigan, at an elevation of 580 feet, to the Mississippi, at 420 feet of elevation, is maintained by dredging and a series of dams with locks.

SPECIAL FEATURES

Since the Eastern Ridges and Lowlands is a showcase for continental glaciation, we emphasized this by selecting the Kettle Interlobate Moraine and the inland lakes of glacial origin as special features. Two of the Great Lakes form the eastern and northern boundaries of this province, so it is also logical to highlight Lake Michigan shorelines, the Door County Peninsula and Pictured Rocks National Lakeshore. These provide an introduction to that geologically active place where land meets water. We also describe the Silurian reefs, since they are a unique aspect of the Paleozoic rocks in this province. Some of these features are included in a special field trip through eastern Wisconsin which is described in our field guide.

Lake Michigan Shorelines, Past and Present

The Lake Michigan shoreline is one of the more fascinating and active geologic laboratories in our region. Stretches of its rocky shores and soft sand beaches are now state parks; and playgrounds for tens of thousands of people each year. Our largest cities also crowd the shoreline, and this results in conflicting needs and attitudes. Probably the only common ground shared by the people that live, work and play along Lake Michigan is the naive desire that the place where land meets water should remain an unvarying line, fixed within the narrow limits of the swash of today's waves. Geology tells us this is not to be, for Lake Michigan has a long history of change, and this will continue no matter where city, county and regional planners draw the lines between land and water on their maps.

The record of many older Lake Michigan shorelines is here for us to read. It is carved in stone and written in clay bluffs, sand beaches and dune fields. We reviewed the geologic history of Lake Michigan, including some basic problems in interpreting it, when we described Pleistocene events in Chapter 2. Let's reflect again upon the higher levels of ancestral Lake Michigan that once inundated areas now occupied by Chicago, Racine, Kenosha, Green Bay, and Escanaba. The terminology we use is summarized in Table VII.

In the Chicago area, the flat deposits that record stands of Glacial Lake Chicago extend inland as much as 15 miles. Recognizable are the 13,500-year-old Glenwood shoreline about 60 feet above present lake level, the younger and lower Calumet at 40 feet, and the Toleston at 20+ feet. Remnants of Glacial Lake Chicago shorelines are also present in a 1.5 miles wide area bordering the Lake Michigan shore from the Illinois state line to north of Racine. However,

TABLE VII
Summary of Ancestral Lakes in the Lake Michigan Basin
(Modified from Willman, 1971)

Lake Stage	Approx. Years Before Present	Approx. Level in Feet Above Lake Michigan (580 ft. above sea level)	Comments
Post-Glacial Levels:			
Algoma	2,000-3,000	10-15	Lowest terrace level above present lake
Nipissing	3,000-4,000	20-25	Similar to Algonquin and Toleston levels
Chippewa	4,000-9,000	Several hundred feet *below* 580 feet	Lowest level
Valderan Levels:			
Algonquin	9,000-10,000	20-25; rises significantly to north	Similar to Nipissing and Toleston levels
Toleston	10,000-11,000	20-25	Similar to Algonquin and Nipissing levels. Last record of Lake Chicago.
Calumet II	11,000-11,500	40	Similar to Calumet I level
Two Creekan Level:			
Bowmanville	11,500-12,000	*Below* 580 feet	
Woodfordian Levels:			
Calumet I	12,000-13,000	40	
Glenwood	13,000-14,000	55-60	Formation of Lake Chicago; highest level recorded

the 20 to 25 foot Toleston terrace level, about 10,000 years old, extends only as far north as Kenosha. The older 40 foot Calumet and 55 to 60 foot Glenwood are traced 3 miles northwest of Wind Point at Racine, until they are terminated by present-day lake erosion. These two levels are occasionally present along the shore between Port Washington and Sheboygan, but here they are somewhat higher in elevation due to postglacial rebound.

Two postglacial beach levels are also present at Shoop Park on Wind Point at Racine, and these represent the Nipissing (3,500 years old) and Algoma (2,500 years old) stages. Here the Nipissing, at 25 feet above present lake level, has essentially reoccupied the former Toleston level, while the Algoma is 10 feet lower. The Algoma beach, about 15 feet above lake level, is well-displayed at Doctors Park in northern Milwaukee County. At Terry Andrae State Park south of Sheboygan, the Algoma level extends inland almost a mile to terminate against a low bluff that exposes sand dunes overlain by pebbly red lake clays that probably represent an older Calumet level of Glacial Lake Chicago. Part of the Algoma lake plain in this general area is blanketed with younger sand dunes.

Beach levels left by Lake Algonquin about 9,000 years ago are present along the Lake Michigan shore north of Two Rivers, Wisconsin. To the south, this level is either eroded away or it coincides with younger or older shorelines. Algonquin beaches are found at increasingly higher levels northward due to postglacial rebound of the land. They rise from about 25 feet above present lake level at Two Rivers, to 40 feet at Sturgeon Bay, 60 feet at Peninsula State Park, and 100 feet at Washington Island. On the Garden Peninsula of Michigan, this shoreline is at the top of the bedrock cliffs, and it climbs to 360 feet near Munising on Lake Superior. In fact, almost all of the Eastern Ridges and Lowlands Province in the Upper Peninsula was submerged at this time.

This dramatic record of Algonquin shoreline tilting (almost 350 feet in 175 miles) is not equaled by younger lake levels. At Two Rivers, the younger Nipissing shoreline coincides with the Algonquin level but it is about 50 feet below it near the end of the Door County Peninsula. The difference in uplift of these two shorelines records how much postglacial rebound occurred over the approximate 6,000 year interval that existed between them. Rebound continues today with the Lake Superior region rising several times faster than the south part of Lake Michigan. This process is slowly but surely forcing the waters of Lake Michigan southward.

General stabilization of outlets and lake level occurred sometime after the 2,500-year-old Algoma level. However, fluctuations of several feet still continue in response to variations in seasonal and climatic cycles. At present time (1976), the lake is relatively high although it has been slowly dropping for several years. Citizens now demanding that the lake level be lowered, will in a few short years insist that something be done to raise it. They are unwilling to comprehend the magnitude and dynamics of this great water system.

The Lake Michigan shoreline, from Chicago to Door County, Wisconsin, is relatively straight with only occasional broad capes. The beaches are narrow and much of the coast consists of steep wave-cut drift or lake clay bluffs that may rise up to 60 feet above the lake. Wave action and gravity-induced erosion have kept the bluffs steep, the slope vegetation youthful, and the beaches narrow (Fig. 6-3). These important processes are discussed in more detail below.

In a few places, there are dune fields along

Figure 6-3. Lake Michigan shoreline erosion between Racine and Kenosha, Wisconsin. Wave-induced erosion of unconsolidated lake deposits, such as these, keeps the Lake Michigan bluffs steep. These landowners are attempting to use groins to develop a beach in the hope of protecting the bluffs from further wave action. On the south (left), it appears that this project is succeeding. However, the result is less beach and increased erosion to the north (right). Milwaukee Journal photo.

the shore and rarely, a low bedrock cliff. Examples are dolomite promontories at Wind Point north of Racine, and Sheboygan Point. Another short expanse of rocky shoreline is present at Harrington Beach State Park north of Port Washington, Wisconsin (Fig. 6-4).

Although there are no large bays along this part of the shore, the mouths of larger rivers flowing into Lake Michigan provide adequate harbors when dredging and breakwater protection is provided. Lake ports include Chicago, Waukegan, Kenosha, Racine, Milwaukee, Port Washington, Sheboygan, Manitowoc, Two Rivers, Kewaunee, Algoma, Marinette-Menominee, and Escanaba.

The shores of the Door and Garden peninsulas, and the rocky islands in the narrow straits between them, are especially inviting. Here, magnificent cliffs of Silurian dolomite rise vertically from the lake only to be interrupted by small, protected bays that contain narrow pebbly beaches (Fig. 6-5). One such bay on the northwest shore of the Garden Peninsula is the site of Michigan's Fayette State Park. This park has additional geologic interest because it contains the restored iron smelting town of Fayette, near the Silurian dolomite quarries where fluxstone was acquired for iron production. The varied shorelines of Door County are described in more detail as a special feature of this province.

Figure 6-4. Lake Michigan shoreline at Harrington Beach State Park, Ozaukee County, Wisconsin. Exposures of Middle Devonian dolomite are present along the lakeshore and at the quarry lake. The buildings are part of the quarry operation which was abandoned in 1925. Wisconsin Department of Natural Resources photo.

Figure 6-5. Headland cliffs of Silurian dolomite alternate with pebbly beaches in parts of the Door and Garden peninsulas, and on many of the islands between them. Rock Island State Park, Wisconsin. Milwaukee Journal photo.

The Stonington Peninsula is across Big Bay de Noc to the west of the Garden Peninsula. This projection is composed of relatively soft Upper Ordovician Maquoketa (Richmond) Shale and it only remains because it is partially protected from wave erosion by the two resistant Silurian peninsulas described above.

The shoreline of Lake Michigan and Green Bay from Escanaba, Michigan to south of Brussels, Wisconsin is generally low, and major rivers (Ford, Menominee, Oconto and Peshtigo) form broad, low deltas. There are very few bluffs along this stretch and beaches are common. There also are several large spits (Little Tail Point and Long Tail Point) that project from the western shoreline for several miles into Green Bay. The general character of the shoreline in this area can be appreciated at Michigan's J.W. Wells State Park, 23 miles north of Menominee.

Fortunately, several other interesting segments of Lake Michigan shore are preserved as state parks. Examples are the adjacent parks of Terry Andrae and J.M. Kohler, near Sheboygan, Wisconsin, which are best known for their partially active dune field. The dunes are 10 to 30 feet high along a quarter-mile wide strip that flanks the beach (Fig. 6-6). This field is a veneer of sand about 45 feet thick that overlies an Algoma terrace cut on Pleistocene drift, and it impounds a marshy lagoonal area that parallels the lake for almost 6 miles. Although these dunes are not as spectacular as those of Indiana Dunes National Lakeshore or Warren Dunes State Park on the eastern

Figure 6-6. Dunes at Terry Andrae State Park near Sheboygan, Wisconsin. This is the older part of the field and the stabilizing effect of vegetation is apparent. Wisconsin Department of Natural Resources photo.

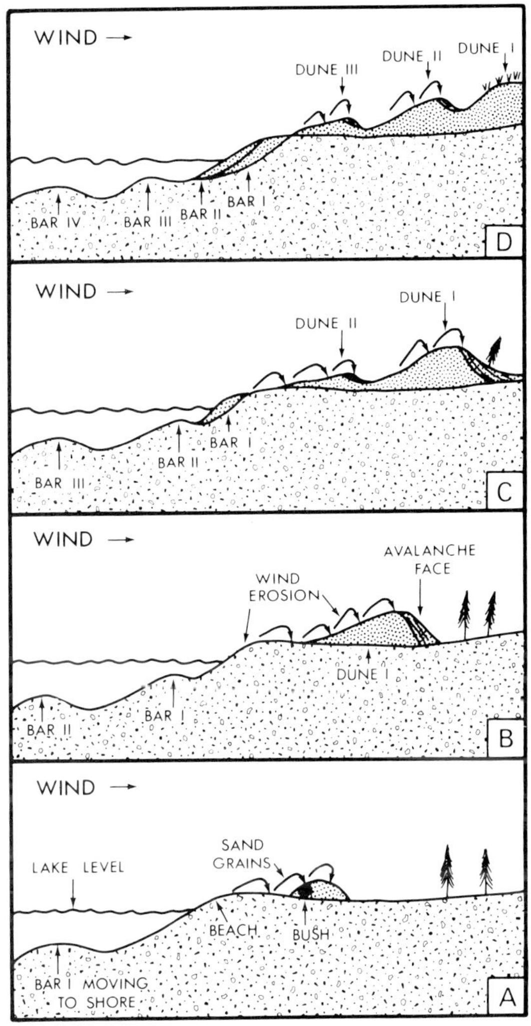

Figure 6-7. Dune formation along an active beach. Offshore bars move onshore under favorable wave conditions to build the beach. Sand from the dry part of the beach is blown inland by onshore winds until it deposits behind an obstacle (A). Once a sand pile accumulates, it provides a wind shadow that causes sand to be deposited on its lee side (A). As this process continues, sand will erode from the windward side of the dune and move up the dune to accumulate just over the edge on the lee face. As it accumulates, it will periodically avalanche down this steep slip face (B). By this process the dune migrates inland while a new dune develops near the beach (C). The younger dunes decrease the sand supply from the beach and block the wind so that vegetation becomes established on the older dune and it is stabilized (D).

shore of Lake Michigan, it is an excellent place to examine beach-dune processes.

The southern and eastern Lake Michigan dune fields receive prevailing onshore winds, but at Terry Andrae-Kohler the dunes only grow during infrequent northeasterly storms. When this happens, dry sand grains are bounced along by the wind. They will stop where they encounter an obstacle such as vegetation or a fence. As the sand pile grows, it creates a wind shadow on the downwind side and a small dune is born (Fig. 6-7). This dune will grow if there is a continuing sand supply, and a strong enough wind to move sand from the front of the dune up to the top. Here, the grains accumulate to avalanche down the lee side. By the combined mechanism of erosion on the windward side and deposition on the lee, the dune will migrate inland. Meanwhile, beach processes operate to renew the sand supply, and a more lakeward dune may form. Ultimately the older, inland dunes have their sand supply cut off by younger dunes, and they are stabilized by grasses and mat-forming shrubs (Fig. 6-7). If plant succession continues, a forest of oak or pine will develop (Fig. 6-6).

Point Beach State Forest, about 40 miles north of Terry Andrae-Kohler, near Two Rivers, Wisconsin, is a constructional shoreline built outward from a broad, low headland. The agricultural land to the west of the state forest is an old shoreline terrace about 25 feet above Lake Michigan, and it was last occupied during the Nipissing stage about 3,000 to 3,500 years ago. As the Nipissing lake level dropped, a spit developed and continued to grow across an embayment along the shoreline (Fig. 6-8). Eventually, the spit joined the shoreline several miles to the south to form a baymouth bar. This bar isolated the bay and a lagoonal lake was created.

This lake eventually filled with sediment and vegetation, while dunes developed on the baymouth bar. A new beach-dune complex formed lakeward of the older shoreline in response to the combined effects of falling lake levels and elevation of the land by glacial rebound. As lake levels continued to decrease, a series of these depositional features formed during periods of lake level stability. Today, each of these beach-dune complexes is a well-defined ridge that can be traced essentially parallel to the old baymouth bar and the present beach (Fig. 6-8). Between the ridges are low, poorly drained areas that represent relatively small but rapid drops in lake level. Ultimately, a broad succession of swales and ridges resulted, where the innermost ridge is the oldest beach-level. This age relationship is reflected in the soil and vegetation, which are progressively more mature on the older ridges.

Once this depositional complex was constructed, the level of the lake stopped falling and erosion began. Much of the geologic work along the Lake Michigan shore is done when northeasterly storm waves attack the beaches and cliffs. This is quite apparent at Point Beach, where wave-induced erosion has cut off the beach ridges at the northern end of this headland. Some of the eroded sand moved south with the longshore drift to build up the present beach on the protected south end of the headland. The result is that Point Beach is migrating southward by beach erosion and deposition. With these conditions in mind, one can speculate on the future of the lodge at the north edge of the park (Fig. 6-8).

Erosional problems associated with the drift and lake clay bluffs exist at many places along the western Lake Michigan shore. Accessible examples in the Milwaukee area include Grant and Big Bay parks. Here, as at many other places along the lake, slump blocks up to several hundred feet wide have moved downslope into the lake, and arcuate erosional scars extend inland for up to 100 feet (Fig. 6-3). Older scars, many now wooded, are visible between fresh cuts. Additional evidence for the youthful state of the lake bluffs is provided by their steepness, and the absence of old trees. Clearly, this process has continued for many years, although recently it has received considerable publicity.

The persistence of shoreline erosion is also documented by measurements, some of which date back over 100 years. Edmund Andrews studied the Lake Michigan shoreline from Chicago to Manitowoc for several years prior to 1870. He determined that the average erosional retreat rate from Chicago to Milwaukee was 6.24 feet per year, while from Milwaukee to Manitowoc it was 4.33 feet per year. In 1877, T.C. Chamberlain reported that between 1836 and 1871, Racine

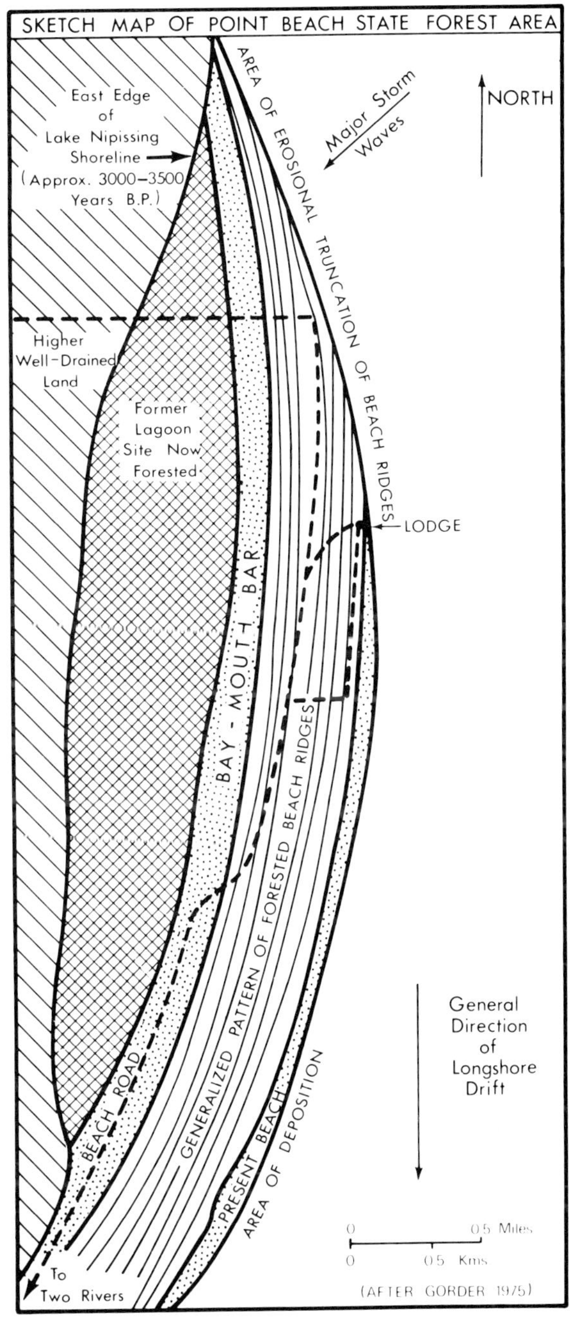

Figure 6-8. Generalized map of Point Beach State Forest area near Two Rivers, Wisconsin. The embayed shoreline of postglacial Lake Nipissing is shown on the left side of the map. A post-Nipissing baymouth bar was constructed across the bay to isolate a lagoonal lake which subsequently was filled with sediment and vegetation. A series of ridges and swales parallel this baymouth bar and the present beach. Each of the ridges represent a stillstand lake level for an interval of time that was sufficient to allow a beach-dune ridge to develop. The intervening swales represent a fairly rapid, but small, drop in lake level followed by the development of another beach-dune complex. The preservation of this remarkable sequence required continually falling water levels. At present, this depositional terrace is eroding at the north end and beach ridges are being truncated. Some of this eroded material is moving southward to build the present beach.

County averaged a shoreline loss of 3.33 feet per year, while it was 2.77 feet per year for Milwaukee County. Lawrence Martin described retreats of 1 to 3 feet per year for the Wisconsin lakeshore during the late 1920s and early 1930s while Lyle Gorder estimated the erosional rate in Manitowoc and Kewaunee counties between 1952 and 1975, at about .5 foot per year. Between 1964 and 1974, the lake rose 5 feet from a record low to about 581 feet. This high water level has once again accelerated the rate of Lake Michigan shoreline erosion.

The geologic record supports these historical measurements. Evidence of glacial shorelines of several ancestral forms of Lake Michigan was once more extensive. Now these records are reduced to narrow remnants scattered along the shore in a few protected areas. At Two Creeks, Wisconsin, geological evidence indicates that erosion has proceeded at about 2 feet per year for several hundred years.

With such overwhelming evidence for an actively eroding shoreline, one would think that a prudent individual would stay clear of the lake margin. Development along the Lake Michigan shore belies such concern, and indicates a lack of respect for the predictability of geologic processes.

We must accept the fact that the shore of Lake Michigan has been a dynamic, shifting boundary since its earliest ancestor formed, and it will continue to change. One can only hope that unstable bluffs will soon be placed under restrictive zoning laws that ultimately would result in the establishment of a buffer zone of lakeside parks. This would protect us from our folly while protecting much-needed recreational land.

The mechanism of shoreline retreat along Lake Michigan is an easy one to understand. It occurs as a continuing series of dramatic earth movements, rather than by a gradual wearing away of the land. During low water stages, normal wave activity builds beaches, and storm waves dissipate their forces on these beaches (Fig. 6-9). In times of high water, the waves direct their erosive energy directly against the relatively soft bluffs (Fig. 6-9). When a cliff is oversteepened by this process, the weight of the overlying material might cause it to slump downward and outward. This occurs most frequently in the spring or fall when the bluff is saturated with ground water. Waves concentrate on a new slump block until it is removed, either offshore or along the shoreline, by longshore drift (Fig. 1-8 and 1-9). On the western coast of Lake Michigan, this longshore drift is generally southward.

Once the slump is gone, waves will attack the bluff again, unless the water level falls. If the water level drops, a beach will be constructed (Fig. 6-9). Even then, unstable bluffs will continue to fail for some time. Ultimately the slumps cease, and lakeshore property owners relax until the next interval of high water, or they sell their property while shoreline erosion is not a matter of immediate concern. Let the buyer beware!

Some shoreline owners try to solve their erosion problems in a variety of ways. A commonly attempted remedy involves the emplacement of a structure that juts outward from the beach into the lake. These are *groins,* and they force the longshore drift into deeper water where its current energy will diminish. This causes sand in transport to be deposited along the groin (Figs. 1-9, 6-3, and 6-10). The hope is that this process will develop and hold a beach to protect the bluffs from wave erosion. Studies along groin sites show that this technique rarely works as planned, and in many instances, erosion above and below the groin area accelerates. Other protection techniques involve offshore breakwaters, covering the

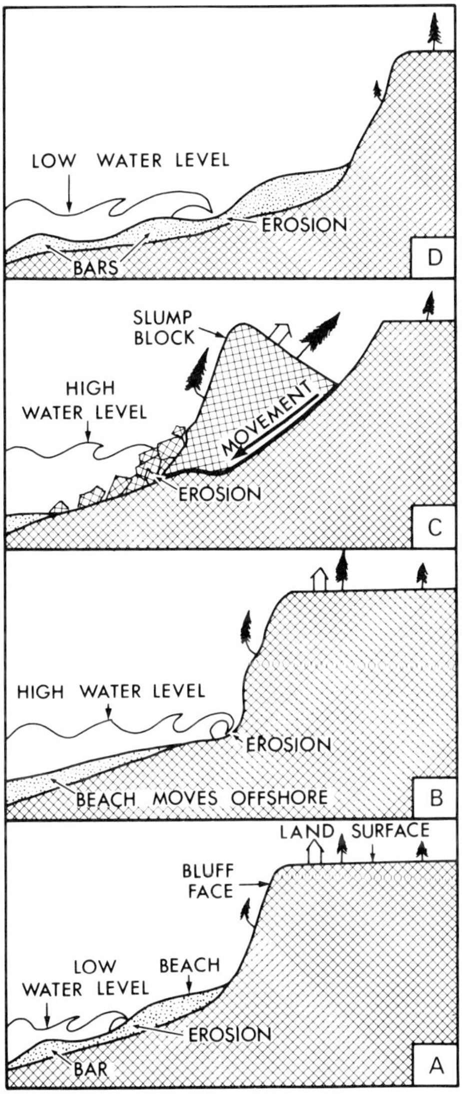

Figure 6-9. Bluff erosion along the Lake Michigan shoreline. At times of low water (A), a beach is developed and it protects the bluff from wave erosion. During high water (B), waves erode the base of the bluff until the oversteepened bluff fails by slumping (C). Waves and currents erode and remove the slump to resume their attack on the bluff unless the water level falls and another beach develops (D).

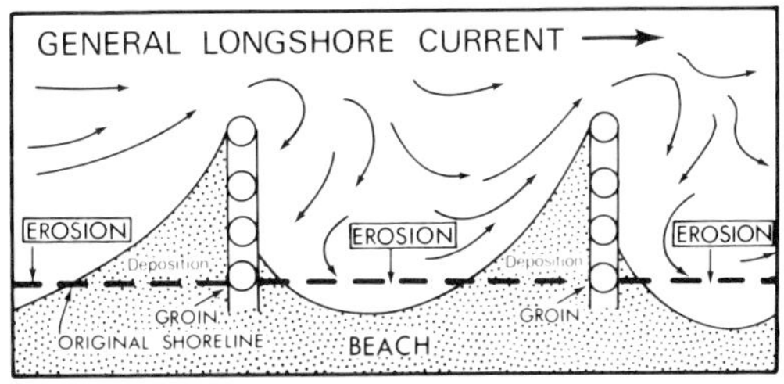

Figure 6-10. Erosion and deposition around groins. The two groins illustrated project outward from the original shoreline as solid barriers that deflect the longshore current into deeper water. This decreases current energy and the sand grains in transport are deposited "upstream" of the groin. Now without a sediment load, the wave-generated longshore current begins to remove sand from behind the groin and to transport it along shore to the right, resulting in erosion of the original shorelines.

shoreline with large blocks of rock (riprap), building seawalls, and constructing offshore "barrier islands." These are all expensive, generally piecemeal, solutions that often create more problems than they solve. In some instances, an individual may solve his problem while destroying a neighboring property.

A good example of the types of problems produced by well-intended shoreline protection programs is provided by Illinois Beach State Park near Zion. At the north end of the park, attempts at erosion control are now focused on maintaining a bathing beach and on saving the lodge. When the Zion Nuclear Reactor was constructed north of the park, a protective jetty was built into the lake. This structure cut off the sand supply moving south with the longshore drift, and erosion accelerated between the reactor and the lodge. Over 100 feet of beach disappeared from this end of the park in a few years. Groins and riprap were rushed into place, and sand was dumped onto the former beach area to retard erosion. However, the south end of the park is not eroding, and it has a wide beach that is presently growing lakeward and southward with sand supplied by erosion to the north. This is the last stretch of natural beach along the lake in Illinois. The present beach flanks a well-developed dune-beach complex similar to that of Point Beach State Forest in Wisconsin. This complex is part of a narrow beach ridge system that began to develop about 3,500 years ago between Waukegan, Illinois and Kenosha, Wisconsin.

Lake Michigan is one of the richest resources in our area. It provides fish, transportation, recreation and a moderating climatic influence. More importantly, it provides a high-quality source of relatively inexpensive water for industrial and domestic use. Unfortunately, it is also the sewer for the surrounding area. In this role, it receives large quantities of runoff from agricultural areas, and the human and industrial waste, some of it untreated, from millions of people and thousands of different industries. More recently, it began receiving hot water from nuclear reactors that are springing up along the lake like fall mushrooms. Only with increased care can Lake Michigan remain a playground and a quality water source, while still serving as a transportation route and a sewer.

Door County Peninsula

Door County is often called Wisconsin's air-conditioned playground, and unlike many Chamber of Commerce promotional themes, it is! Situated between the air conditioners of Green Bay and Lake Michigan, it is a narrow (7 to 20 miles wide) Silurian bedrock peninsula that rises nearly 200 feet above present lake level. It is a picturesque vacation spot comparable to the rock-bound coasts of New England. Glacial drift is thin or absent, and silver-gray dolomite is evident in bold cliffs, roadcuts, and as fence rows bounding pastures and cultivated fields of tidy farms. Cool, dim cedar glades line the moist nearshore margins while birch, aspen and pine blanket the uncleared uplands.

The waters of Green Bay and Lake Michigan do much more than keep tourists cool. It was commercial fishing that first brought the Scandinavian settlers here, and some of this still exists. Presently, sport fishing is now more important. Shipbuilding, an early industry in this area, prospers at Sturgeon Bay. Agriculture is also important, although the thin, stony soils are not especially fertile, and much of the land remains forested or is used for pasture. The climate is ideal for fruit trees, however, and the growing season here is as long as that of southern Wisconsin, due to the moderating influence of the water. The winter-chilled lake holds spring back long enough to prevent cherry and apple orchards on the upland surface from blooming before the frost danger is over, while the summer-warmed waters hold the frost back until the fruit matures.

The western, or Green Bay side of the peninsula is the cliff face of the Niagaran cuesta and it is characterized by towering dolomite headlands and bluffs. The occasional reentrants form bays with sand or locally-derived cobble beaches at their heads, and many were sites of former fishing villages. Now they are busy tourist towns during the summer, but the maritime flavor remains. In contrast, the eastern or Lake Michigan shoreline has long stretches of sandy beach, dune fields, and only occasional exposures of bedrock. This side of Door County is the backslope of the Silurian escarpment as it dips southeastward under the lake into the Michigan basin.

Door County received its name from the infamous channel between the main part of the peninsula and Washington Island. This was the Porte des Morts (Death's Door) passage of the early French explorers; so named because of the unpredictable strong currents and fierce winds which would drive their canoes against the rocky shore. LaSalle's Griffon, the first sailing vessel on the Great Lakes, was supposedly shipwrecked here in 1679 with a load of furs. An inland sea "Flying Dutchman" legend grew from this disaster and a few individuals still report seeing the ghost of the Griffon scudding wildly ahead of gale winds. Since 1679, hundreds of other vessels were wrecked in this wicked passage. Now a ferry carries tourists, supplies, and mail to Washington Island across this strait from one of two alternate docking sites, depending upon the wind direction.

We previously described how the present-day Sturgeon Bay Canal was partially carved by a preglacial river. Port des Morts apparently was also cut by a preglacial stream flowing off the Northern Highland, as were the gaps between all the islands scattered along the 30 mile stretch between the end of this peninsula and the Garden Peninsula of Michigan. These isolated segments of the Silurian escarpment were later inundated by the rising waters of Lake Michigan (Fig. 6-5). Sixteen of these islands are now under consideration as a national lakeshore area.

Shallower gaps were made in the bedrock of the peninsula by streams flowing down the dip slope of the cuesta. These gaps head near embayments on the Green Bay side, but they are not deep enough to be flooded by the lake. Small streams, which drain southeasterly across the peninsula to enter Lake Michigan, presently occupy most of these cuts. Rising water levels flooded the mouths of these streams to form protected bays, which later became the sites of small settlements. Some of these bays were modified by the deposition of sandbars across their outlets. In several cases, this continued until they were isolated as coastal lakes. Examples are Clark, Kangaroo, and Europe lakes.

The few islands in Green Bay are outliers of the Silurian cuesta. Chambers and Horseshoe islands, and the Strawberry and Sister groups somehow escaped the glacial planing which straightened the escarpment. The low-lying Sister Islands are now a protected breeding ground for herring gulls, and terns, while Horseshoe is part of Peninsula State Park.

Shorelines of abandoned, higher levels of ancestral Lake Michigan are recognizable in many places in Door County. Wave-cut cliffs, sea caves, and beach ridges, now lifted by glacial rebound to more than 60 feet above present lake level, were formed by erosion and deposition during the stands of lakes Algonquin and Nipissing (Fig. 6-11).

Like Point Beach and Illinois Beach, The Ridges Sanctuary just north of Baileys Harbor is a succession of abandoned beach-dune ridges separated by intervening swales. Each ridge formed as a beach with dunes at the head of a large bay (bayhead beach) rather than along a headland as at Point Beach. Preservation of these beach-dune features required that the land rise relative to the water, and this resulted from a combination of postglacial rebound of this area and decreasing water level. The Ridges complex has 16 remarkably parallel, sandy ridges that gracefully curve to reflect the shape of the original bay. Each is a record of a stillstand in lake level that persisted long enough for a beach to develop. Then, the land-water relationship shifted lakeward where a younger

Figure 6-11. Cliff of well-bedded Silurian dolomite at Eagle Bluff, Peninsula State Park, Door County, Wisconsin. The dark area about 30 feet up on the cliff face is a sea cave that was carved during a higher, postglacial level of Lake Michigan. Wisconsin Department of Natural Resources photo.

beach formed. This process is continuing today since a modern beach-dune is developing in the bay across the road from The Ridges, and it is essentially parallel to the older ones.

The Ridges is also a botanical paradise where a rich assortment of wild flowers grows in alternating wet and dry environments; for the low areas between the well-drained ridges are either water-filled sloughs or moist meadows. Moving inland, each ridge is progressively older with a more mature soil, an older and better established plant community, and more shade. Many plant species are boreal in origin, and all are adapted to life in either dry sand or cold, acid bogs. The diversity of the flora is striking, and it includes two-thirds of Wisconsin's native orchid varieties.

The Ridges remained as virgin land because the heart of the sanctuary was established as a federal range light, used to guide sailing ships into Baileys Harbor; the earliest settlement on the peninsula. The original 40 acre facility consisted of a lighthouse near-shore and a range light inland. When the two lights were lined up, the ship's master could sail safely into the harbor.

In 1936, the government gave the land to Door County for a park and campground. This potential use involved timber cutting, drainage and leveling. Many individuals tried to impress county officials with the need to preserve this unusual tract of land, but they failed. As work started, a resolute group of women from Baileys Harbor stood in the way of the bulldozers and halted "progress." County officials yielded in the face of this protest, and to pressure from other quarters, and a wild flower sanctuary was established. At the present time, The Ridges Sanctuary encompasses more than 900 acres of good geology and botany.

Because of its varied scenery, Door County is the site of five state parks. Potawatomi State Park, 3 miles west of Sturgeon Bay, is dominated by Government Hill, a bedrock bluff that rises 150 feet above the waters of Green Bay. The name for this hill acknowledges that the United States once held these lands and quarried Silurian dolomite. An overview of Green Bay, with the cuesta front exposed in huge quarries across the Sturgeon Bay canal, is available from an observation tower on Government Hill.

Peninsula, located between Fish Creek and Ephraim, is another park overlooking Green Bay. Here, bold cliffs of Silurian dolomite are cut by reentrants that head in sandy beaches. In the Fish Creek embayment southwest of the park, 8 postglacial abandoned beaches are recognized above the present beach. The highest is the Algonquin, now 60 feet above lake level. The Nipissing is 21 feet above the lake and remnants of two lower levels are present below this. The high point in the park is Eagle Bluff, which rises 180 feet above lake level to provide a magnificent vista. Below this bluff is Horseshoe Island, shaped to fit its name. One of the more dramatic records of former lake levels is the sea cave carved in the face of Eagle Bluff, 30 feet above the water (Fig. 6-11). This feature was cut thousands of years ago by the same processes of wave erosion that are now operating at Cave Point, about 17 miles south on the Lake Michigan side of the peninsula (Fig. 2-23). If you make this comparison yourself, we think you will agree that it is a fine illustration of the Law of Uniformitarianism.

Rock Island is an undeveloped offshore park less than a mile northeast of Washington Island (Fig. 6-5). Much of the shore consists of wave-cut cliffs, but one bay with a beach is present. Newport, a few miles east of Ellison Bay along the shores of Europe Bay, is also largely undeveloped. This park includes the site of the pioneer lumbering community of Newport which became a ghost town in the early 1900s after all the timber on the peninsula was cut. The 7 mile shoreline of this park consists of alternating rocky and sandy stretches. Above the present beach, Algonquin and Nipissing shorelines are preserved.

Whitefish Dunes, southwest of Jacksonport, is one of Wisconsin's newest parks, and it typifies the character of the Lake Michigan shoreline. Much of the beach here is wide and sandy. Behind the shoreline is a large dune field now stabilized by vegetation, with many dunes that rise 50 feet above interdune areas. The beach and dune field are both integral parts of a baymouth bar that grew southward to dam an original irregular bay, and isolate the inner part as Clark Lake. This park includes Cave Point, a promontory of Silurian dolomite continually under attack by waves, that anchors the north end of the baymouth bar that formed Clark Lake. The cliffs on this point are 30 to 50 feet high and they are carved into ledges, arches, and sea caves. If you should be fortunate enough to catch storm waves directed against this headland, you will never forget the erosive power of wave energy along the Lake Michigan shoreline.

Pictured Rocks National Lakeshore

The Pictured Rocks are imposing wave-cut sandstone cliffs that extend northeastward along the south shore of Lake Superior from Munising toward Au Sable Point. Part of this stretch is designated as a national lakeshore, which guarantees that these scenic cliffs and sandy beaches will remain available to the public. The rocks here are tan and reddish-stained sandstones of the gently southeast-dipping Upper Cambrian Munising Formation. At Pictured Rocks, the Munising is composed of a lower member (Chapel Rock Sandstone) near lake level, and an overlying Miner's Castle Sandstone Member. The Chapel Rock contains materials eroded from nearby highland areas to the south, and deposited in a northerly arm of the transgressing Late Cambrian sea. The younger Miner's Castle is also a marine deposit, but it was deposited during a more widespread Late Cambrian transgression. It contains more quartz grains, and the sand is rounder and better sorted.

The sandy dolomite of the Lower Ordovician Au Train Formation overlies the Munising. This formation was deposited at the top of the Late Cambrian-Early Ordovician transgressive cycle in this area. The resistant Au Train forms an escarpment essentially parallel to Lake Superior between Marquette and Sault Ste. Marie, Michigan. Along this escarpment, many northward flowing streams cascade over the fall line enroute to Lake Superior. One of these waterfalls, Miner's Falls, is within the national lakeshore (Fig. 2-19). The Au Train Formation is also the cap rock that protects the more easily eroded sandstones of the Pictured Rocks (Figs. 2-4 and 6-12).[1]

The national lakeshore contains wave-sculptured features known as Miner's Castle and Chapel Rock, which provide convincing evidence of high postglacial lake levels (Fig. 6-13). Two-thirds of the way up the promontory cliff face of Chapel Rock is a cave which was carved when Lake Superior stood at a higher postglacial level. By subsequent erosion, this abandoned sea cave became the arch we see today. Miner's Castle juts outward from the Pictured Rocks cliff as an intricately carved feature that displays current bedding within the Munising Formation. On the top of this feature there are two, high and dry, sea stacks that were also developed during a higher stand of Lake Superior. These are now the turrets of Miner's Castle. (Fig. 6-13).

The present elevations of Chapel Rock sea cave and the Miner's Castle sea stacks do not reflect the true elevations of former post-glacial lake levels. These features, as part of the southern Lake Superior area, started rising by rebound as the Pleistocene ice retreated and this process continues today.

Kettle Interlobate Moraine

The Kettle Interlobate Moraine is a linear complex of glacial deposits and lakes that extends through eastern Wisconsin from northern Manitowoc County southwestward to west-central Walworth County (Fig. 2-28). Its irregular, relatively high-standing topography varies in width from 1 to 30 miles and rises several hundred feet above subdued ground moraine surfaces which flank it on the east and west (Figs. 6-1 and 6-2). This impressive feature didn't fail to attract the attention of both early settlers and geologists. William Alden, who described the geology of southeastern Wisconsin in 1918, recognized the interlobate moraine to be the "master topographic feature of the whole series of glacial deposits in eastern Wisconsin." This description was appropriate since it marks the merged terminal moraines of the Green Bay and Lake Michigan lobes of the Woodfordian glacial advance.

When these two lobes first approached each other, their thickness was not great. As they advanced, however, the thickness of the ice along their juncture increased from sev-

1. Although Pictured Rocks are off the edge of Figure 2-4, the relations of the Cambrian and Ordovician along the south shore of Lake Superior at the map's edge continue eastward into this area.

Figure 6-12. Spray Falls cascades into Lake Superior over the Upper Cambrian Munising Sandstone along the Pictured Rocks National Lakeshore, Alger County, Michigan. Mottled red and tan weathering colors along this sandstone cliff provide the name for this scenic shoreline which is best viewed from tour boats that operate from Munising during the summer. Michigan Tourist Council photo.

Figure 6-13. Miner's Castle, a wave-carved projection along the Pictured Rocks National Lakeshore, Alger County, Michigan. The less resistant sandstone unit at water level is the Chapel Rock Sandstone while the rest of this feature is composed of Miner's Castle Sandstone; both are members of the Upper Cambrian Munising Formation. The twin turrets of the castle were formed as sea stacks during a higher, postglacial level of Lake Superior. Michigan Tourist Council photo.

eral hundred feet to several thousand feet, and discontinuous, essentially parallel, morainal ridge systems were formed. As the ice stagnated and began to thin, sediment-laden meltwater flowed down cracks and holes in the ice to provide a source for subglacial streams. Meltwater also cascaded over the ice surface into the widening abyss between the tongues, where large chunks of ice were buried in the outwash sands and gravels. A complex of crevasse fills, moulin kames, eskers, outwash, and kettles resulted, and this was superimposed on the morainal ridge systems (Figs. 2-33, 2-34, and 2-35).

Glacial spillways, either abandoned or now occupied by undersized streams, mark major meltwater drainage channels across parts of the interlobate moraine. Terraces paralleling the sides of these valleys record flood levels of swollen, glacial streams. From the observation tower at Lapham Peak, near Delafield in the southern part of the Kettle Interlobate Moraine, the southwest trending Genesee spillway with its outwash terraces is visible. This vantage point also provides an overview of typical Kettle Moraine topography in an area where the interlobate moraine coincides with the edge of the underlying Silurian escarpment.

Another spectacular view over this scenic countryside is from the church steeple at the Carmelite Monastery on Holy Hill, south of Hartford in Washington County. Holy Hill is a moulin kame perched on the interlobate moraine and it is the highest point in southeastern Wisconsin (1,361 feet above sea level) (Fig. 6-2). From here, one can look outward from the western slope of the interlobate moraine into the drumlin-studded lowland once occupied by the Green Bay lobe.

Many classic kames and eskers within the Kettle Interlobate Moraine were destroyed in the quest for convenient supplies of sand and gravel. In the late 1930s, however, Wisconsin began to acquire parts of this glacial masterpiece and to incorporate them into the Kettle Moraine State Forest. Consequently, hundreds of magnificent glacial features are preserved within both the north and south units of this forest. In the southern unit near Eagle, the Wisconsin State Historical Society recently completed a Bicentennial project called Old World Wisconsin. This open-air museum is dedicated to the many diverse ethnic groups which helped to settle this region. Much of the northern unit is within the Ice Age National Scientific Reserve and a new state park at Pike Lake near Hartford was created as part of this project (Fig. 2-36).

Inland Lakes

The Eastern Ridges and Lowlands, like the Northern Highland, is a land of lakes which add beauty and variety to this region. They are the frosting on our glacial cake, for if a single explanation for them were needed, it would be glacial deposition. Preglacial stream patterns were destroyed during glaciation, leaving the area poorly drained. Some of the earlier drainages were plugged by morainal dams, stagnant blocks of ice left kettle holes behind in the drift or outwash, and the entire province was generously treated to an irregular distribution of glacial debris resulting in a maze of depressions and ridges.

Six general clusters of lakes are present in this province. These are: the Kettle Moraine lakes of Waukesha and Washington counties, Wisconsin; the southeastern Wisconsin lakes of Walworth and Racine counties, the northeastern Illinois Fox Lake area, the Four Lakes of Madison, Wisconsin; Horicon Marsh and the Rock River lakes of Wisconsin, and Lake Winnebago. Lakes in the Kettle Interlobate Moraine are primarily kettle holes in till or in outwash areas within the morainal complex (Fig. 2-35). However, Pewaukee Lake is an exception since it formed by morainal damming of a preglacial drainage.

The lakes of southeastern Wisconsin and northeastern Illinois are also related to major moraines. Those in western Walworth County, Wisconsin (Lakes Geneva, Delavan, Whitewater, etc.) are related to the Woodfordian terminal moraine, while those farther east in Wisconsin and the Illinois lakes area, are related to well-developed Woodfordian recessional moraines. Most are kettle lakes, but the large and somewhat linear basins of lakes Geneva and Delavan result from the damming of preglacial river valleys. Big Foot Beach State Park (Lake Geneva) and Chain of Lakes State Park to the south in Illinois provide opportunities to enjoy the glacial lakes of this region.

The Four Lakes of Madison are the product of glacial erosion and deposition along the Yahara River valley. A similar combination of glacial activities along the Rock River valley accounts for Horicon Marsh and Lake Koshkonong. Lake Sinissippi, south of Horicon Marsh along the Rock, is an artificial impoundment.

Lake Winnebago, with an area of 215 square miles and a maximum depth of 21 feet, is the largest inland lake in our area. It sprawls for 28 miles along the west edge of the Silurian escarpment. During the Pleistocene, its basin was excavated from relatively soft Ordovician bedrock on the drainage divide between the south-flowing Rock and northeastward-flowing Fox.

Geologically speaking, all lakes are temporary features. Processes of sedimentation and the invasion of vegetation progressively fill the lake basin (Fig. 6-14). As this continues, a bog or marsh develops which slowly evolves into a poorly drained meadow. All stages of lake extinction are present within the Eastern Ridges and Lowlands, but they are especially easy to observe in the smaller kettle lakes. The regime of some lakes ends when their morainal dams are breached by outflowing streams.

Our way of life plays an increasingly important role in accelerating the death of our

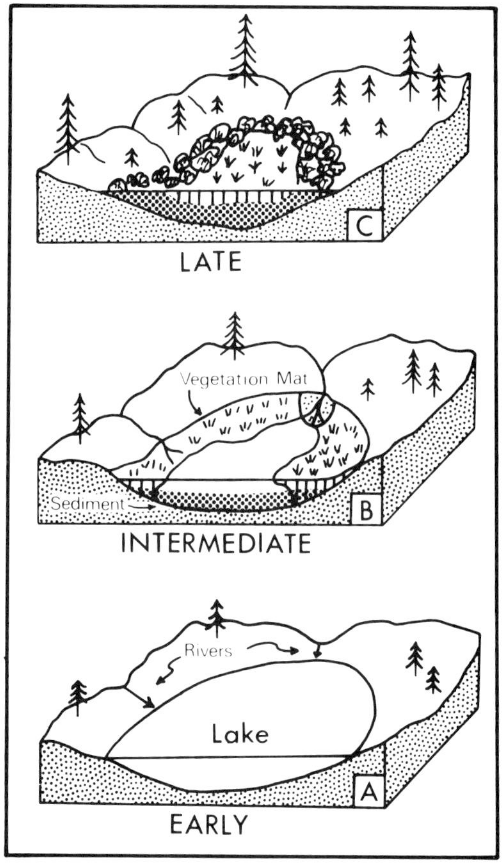

Figure 6-14. Block diagrams illustrating the progressive destruction of a lake. In (A) the lake occupies a depression newly created by one of many geological processes. Drainage into the lake contributes sediment, and vegetation develops in shallow water (B). The processes continue until the lake is shallow enough to support vegetation everywhere and open water disappears (C). A swamp now exists in place of the lake, and this will continue to fill in.

lakes. Water tables are lowered by excessive use of ground water supplies by industry and agriculture. Modern agriculture and major construction projects result in large areas of unprotected soil which may be transported into lake basins. Pollutants of many kinds are also concentrating in our lakes. One of the commonest of these is fertilizer, which washes from agricultural lands. Fertilizer enriches the lake water with nutrients which accelerate the development of vegetation. Restraints on many of these destructive processes now exist, but additional efforts are needed. Our lakes die of natural causes soon enough without additional help from us.

Silurian Reefs

By the Middle Silurian, seas spread throughout our region and covered the Kankakee and Wisconsin arches along the western and southern margins of the Michigan basin. The widespread extent of this sea served to isolate most of our area from sources of land-derived sediment on the Canadian shield and from an orogenic belt far to the east. The water over the arches was relatively shallow, clear and warm, with sufficient wave and current agitation to provide ideal sites where communal, calcite-secreting animals and plants such as algae, corals and stromatoporoids (extinct relatives of the corals) formed reefs. Similar conditions existed on the other flanks of the Michigan basin at this time, so it was encircled with growing reefs (Fig. 6-15).

That part of a reef which was bound rigidly into a wave-resistant structure by reef-forming animals and plants is called a reef core. The Silurian reefs grew upward from the sea bottom from a few feet to almost 100 feet to reach the agitated surface water. Surrounding the reef core, and transitional into it, are flank deposits that dip radially away at angles of 10 to 50 degrees (Fig. 6-16). This

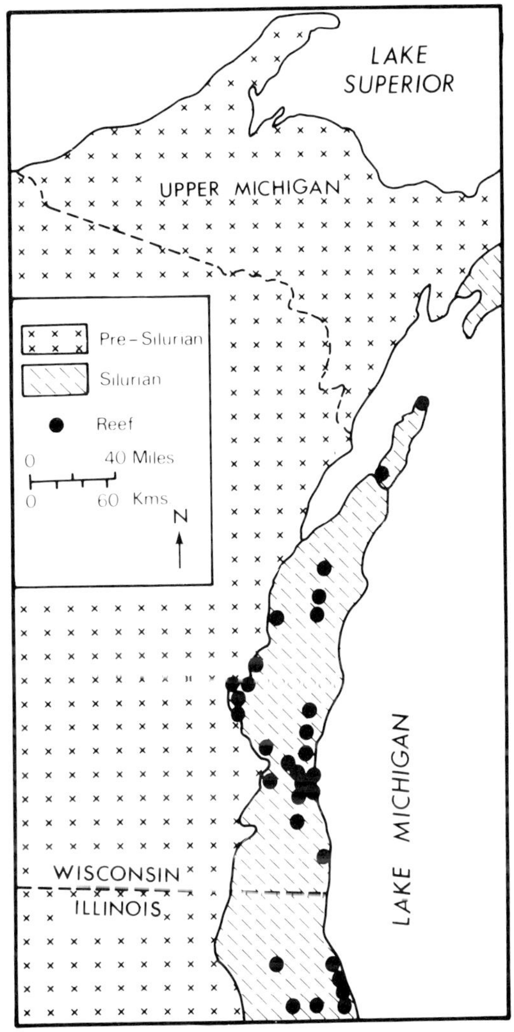

Figure 6-15. Distribution of Silurian rocks and the location of known Silurian reefs within the Eastern Ridges and Lowlands Province. The area shown is along the west flank of the Michigan basin. Modified from Wolosin, 1972.

part of a reef complex is the most spectacular thing we see preserved today, since the core is almost always a featureless rock mass. The reef flank consists of thick, crudely bedded rocks which contain fossil debris,

Figure 6-16. Dipping reef flank deposits at Quarry Lake Park, Racine, Wisconsin. This reef complex can be examined from the ice during the winter.

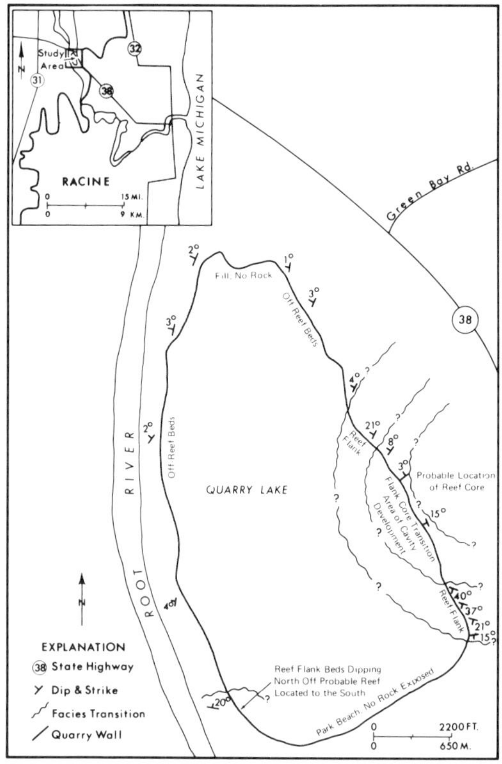

Figure 6-17. Map of Quarry Lake and Silurian reef-bearing rocks at Quarry Lake Park, Racine, Wisconsin. A reef complex in the Silurian Racine Dolomite is well exposed on the southeast wall of this abandoned quarry. The rocks mapped as flank-core transition may include parts of the reef core.

(commonly crinoids) and coarse, angular fragments of reef core rubble (breccia).

Flank deposits grade outward into dense, flat-lying, thin-bedded, inter-reef rocks. These rocks are composed of fine organic debris, largely derived from the reef and transported into deeper, quieter water by waves and currents. Fossils are rare, and this suggests that the inter-reef flats were sterile environments. Much of the Silurian rock in our area represents interreef deposits. This dense rock with its thin, even bedding is an ideal building stone, and many quarries now exploit this resource.

The Silurian rock in our area was dolomitized, and this altered and obscured many of the depositional features associated with reef complexes. Most cores are structureless features, with little evidence remaining to prove their organic origin. It is usually the steeply-dipping flank beds that establish the location of a reef by interrupting the thin, horizontal bedding of the inter-reef deposits.

There were many reef exposures within this province in the past but few remain, and most of these are not accessible to the public (Fig. 6-15). The massive cores resisted glacial erosion and often protruded through the Pleistocene drift as bedrock knobs. Such exposures were favored quarry sites in the early days of urbanization and some, like the

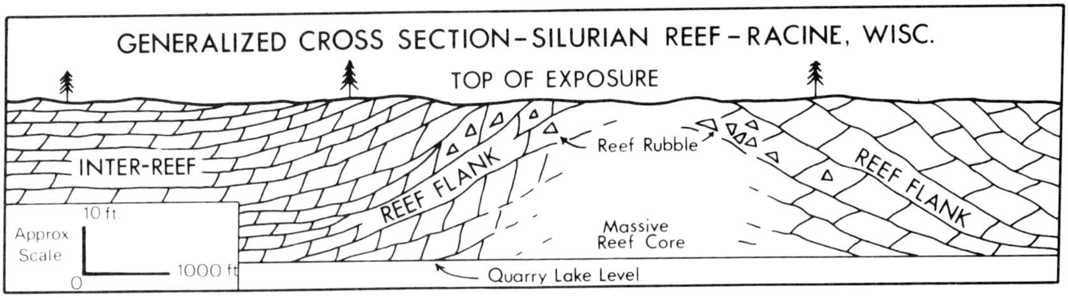

Figure 6-18. Diagrammatic cross section of the Silurian reef complex at Quarry Lake Park, Racine, Wisconsin. The relations between flat-lying, thin-bedded, inter-reef rock; steeply dipping, thick-bedded reef flank deposits; and the massive flank-core transition are shown.

Thornton quarry south of Chicago, are still active. After quarrying, most of the reef localities were filled with refuse and later used for building sites. Most of the Silurian fossil collections now held by museums and collectors came from fossiliferous reefs exposed in these old quarries.

The best public Silurian reef locality within this province is at Quarry Lake Park in Racine, Wisconsin (Fig. 6-17). This reef is within the Racine Formation, and it can be studied by walking on the ice of the lake during the winter. The exposure is such that the inter-reef beds, reef flank and part of the core are available for examination (Fig. 6-18). It is quite an experience to stand on the wind-swept ice of Quarry Lake during a harsh Wisconsin winter and look up at a coralline reef that grew in a tropic sea more than 400 million years ago.

Chapter 7

MINERALS, ROCKS AND FOSSILS

Minerals are solid, naturally occurring inorganic elements or compounds with a definite chemical composition and an internal crystalline structure. Most rocks are mixtures of minerals, but one mineral is often dominant. Fossils are present in most sedimentary rock types, and occasionally they constitute all, or most of the rock. Collecting minerals, rocks, and fossils is one of the strongest human traits, if measured by the contents of the pockets of the young.

This chapter provides an introduction for the collector, but it does not contain specific information on the identification of minerals, rocks and fossils. There are many excellent books on these subjects that can be consulted. We also recommend a visit to one or more of the geologic museums in this region where collections may be studied.

All collectors should use good manners and common sense. Trespassing is unforgivable and it will spoil future access to a locality for everyone. Rockfalls in quarries and roadcuts are a constant hazard; as are rock or metal chips generated by hammering. Hard hats, good boots, and safety glasses are wise investments.

Most collectors take their samples in moderation and with great care. To them we say, "Thank you!" Unfortunately, a few try to clean out an entire locality, and some feel compelled to destroy good specimens that are too large or too difficult to collect. Other forms of geological vandalism that have increased in recent years include the use of geologic hammers to mar scientifically significant rock exposures. In some instances this is done by student groups while under the "supervision" of an earth scientist. The individual who paints geologic information directly on rock surfaces is another source of distress. It is probably foolish to be offended by this well-intentioned action after we see what "John loves Suzy" addicts do with their spray paint cans. However, most of us prefer our rocks unpainted by anyone.

MINERALS

In Chapter 1, we discussed rock terminology and listed 17 minerals that are important in the identification of rock types. These are the common rock-forming minerals, and they are almost entirely composed of only 8 chemical elements. These elements, which constitute 98% of the earth's crust, are oxygen, silica, aluminum, iron, calcium, magnesium, sodium, and potassium. Some of the rock-forming minerals are also economic resources of this region. Even though these minerals are common, they are also prized by collectors when found as large, well-formed crystals. Although very few minerals make up most of the earth's crust, there are more than 2,000 other varieties. Among these relatively rare minerals are those that provide most of our metals.

Human interest in minerals is as old as the species. We became dependent upon them for our tools of agriculture, industry, and war when the first flint hoe, knife, and projectile was fashioned. For thousands of years we used precious metals and gems for personal decoration and as a medium for trade. Viking sailors even used magnetic and light polarizing varieties to navigate across uncharted seas to the New World. Today we still seek minerals for tools, trade, and beauty. Those with commercial value are called economic minerals, and the logical place to collect them is where they are, or were, mined. Minerals with no commercial value, but still of interest to a collector, are often found in association with economic minerals. For example, calcite, marcasite and pyrite crystals associated with the ores in the Upper Mississippi River Lead-Zinc District are prized specimens in many local collections.

A great variety of minerals is present throughout our region and the following list cites only a few of the many collecting localities.

Agates—Pebble beaches on the Lake Superior and Lake Michigan shorelines, but especially along the northwest shore of the Keweenaw Peninsula, Michigan and the Lake Superior shore of Minnesota. Associated with beach agates are epidote, thompsonite, prehnite, analcite, and chlorastrolite. Agates are also found in glacial drift throughout our area.

Asbestos—Marinette County, Wisconsin and near the old Ropes Gold mine north of Ishpeming, Michigan.

Beryl—Near Tigerton (Shawano County), Wisconsin and in pegmatite near Republic, Michigan.

Calcite—Crystalline calcite may develop wherever cracks or cavities (vugs) are present in Paleozoic carbonate rocks. It is also common in lead-zinc, copper and iron mining areas.

Chert—Numerous varieties, including milky, chalcedony, yellow and red jasper, banded chert, flint, etc., are common in Pleistocene gravel deposits throughout the region. Attractive, banded chert nodules may be found in Silurian dolomite.

Copper Minerals—Native copper, copper sulfides, silicates, oxides and carbonates are present in northwestern Wisconsin, northeast of Duluth, Minnesota along the Lake Superior shore, and especially on the Keweenaw Peninsula of Michigan. (See Copper Country in Chapter 3.) The copper minerals are present within cavities and cracks in lava, and within conglomerate interbedded with lava (Fig. 7-1). Associated are the minerals found with agate as described above, plus datolite, apophyllite and analcite. Native copper is also found occasionally in Wisconsinan glacial drift in Wisconsin and Illinois.

Iron Minerals—See Iron Ranges in Chapter 3. In addition to most of the ores of iron (hematite, limonite, magnetite, martite, jaspilite, etc.), barite and several man-

Figure 7-1. Native copper from the Cliff mine, Keweenaw County, Michigan. The irregularity of this sample records the shape of the fissure in which the copper was deposited. The long dimension of the sample is 7 inches. Photo by Frank C. Charnon.

ganese minerals are present on the numerous dumps in iron mining areas. Limonite (goethite) masses are present at the old Iron Hill mining area 2 miles north of Waukon (Allamakee County), Iowa.

Garnet—On mine dumps near Michigamme and Republic, Michigan. Crystals up to one inch across are occasionally found.

Kyanite—Near Powell (Iron County), Wisconsin.

Lead and Zinc Ores—See Lead-Zinc District in Chapter 5. Concentrations of mine dumps in Wisconsin are located near Cobb, Shullsburg, Mineral Point, Dodgeville and Linden. Dumps are also located around Galena and Elizabeth, Illinois, and near Dubuque, and in northern Allamakee County, Iowa. Common minerals are galena and sphalerite associated with pyrite, marcasite, smithsonite, and calcite (Fig. 7-2).

Molybdenite—West of Mountain (Marinette County), Wisconsin; northwest of Ishpeming, Michigan at the old Ropes Gold mine, and from mine dumps at Republic, Michigan.

Pyrite—Common in Paleozoic carbonate rocks, and on the dumps in both the iron and lead-zinc mining areas.

Quartz—Crystals are found in waste dumps of most iron mines and the Keweenaw copper mines. At the Isle Royale mine south of Houghton, prehnite and epidote are associated with the quartz. Another well-known locality is just north of Rock Springs (Sauk County), Wisconsin where small crystals formed in fractured Baraboo Quartzite.

Staurolite—South side of Lake Michigamme, Michigan, and mine dumps at Michigamme and Beacon.

Tourmaline—Within quartz on mine dumps near Michigamme, Michigan; from pegmatite near Republic, Michigan; and mine dumps at Beacon, Michigan.

Figure 7-2. Cubic galena covered with tiny crystals of pyrite; Galena, Illinois. The largest cube is 1.5 inches across. Sample courtesy of The University of Wisconsin-Milwaukee Greene Museum; photo by Frank C. Charnon.

Tremolite—Marble quarry near Felch, Michigan.

ROCKS

The geology of this region is ideal for an individual who wishes to examine and collect all of the igneous, metamorphic and sedimentary rock types described in Chapter 1. The geologic map (Fig. 2-4) and Chapter 2 provide a general guide to areas where rocks of various types occur. Pleistocene gravel deposits and pebble beaches along the shores of Lake Michigan and Lake Superior are also excellent collecting localities, for they contain rock types that were transported southward by the glaciers. Many specimens from a beach are especially prized when they are rounded and polished. If you collect in a gravel pit, don't forget to keep your eyes open for one of those Wisconsin diamonds. You never know!

Igneous rocks are present throughout the Northern Highland. Varieties of Wisconsin granite may be collected from quarries or outcrops in the vicinity of Montello, Wausau, Redgranite, Berlin, Merrill, and Amberg. Somewhat unusual Wisconsin rock types are anorthosite near Tigerton, syenite just west of Wausau, and gabbro at Mellen. Basaltic and rhyolitic lava are well exposed along the Keweenaw Peninsula in northern Michigan. In Minnesota, gabbro is the bedrock at Duluth, and basaltic lavas are exposed along Lake Superior north of the city.

All the common sedimentary rock types are easily located and widely dispersed within our region, although conglomerate is less common. Conglomerate collecting localities include Pleistocene gravel pits where the gravel may be locally cemented with calcium carbonate, and the basal Paleozoic rocks adjacent to the Baraboo Quartzite in Sauk County, Wisconsin. Several relatively unusual sedimentary rock types are also present in our region. One is glauconite-bearing Upper Cambrian sandstone, which is common in central and western Wisconsin. Another is the oolitic hematite of the Ordovician Neda Formation of eastern Wisconsin and northern Illinois.

The Northern Highland is the best area for collecting metamorphic rocks. Gneiss and schist are intermittently exposed along the Wisconsin River from south of Merrill to Nekoosa. Slate and quartzite are common rock types in the iron ranges of Wisconsin and Michigan. Specific exposures of quartzite, sometimes associated with schist or phyllite, were described as a special feature of the Northern Highland in Chapter 3. Quartzite also occurs in the Baraboo Range of central Wisconsin. Marble quarries are located about 25 miles south of Ashland, Wisconsin, and in northern Michigan near Felch. Amphibolites are also present at Felch.

FOSSILS

In our region today, the land and water support a great variety of plant and animal life, but the record of little of it will remain very long after death. Similar conditions existed in the geologic past, so fossils, which are any record of former life, are relatively uncommon. In rare instances, the fossilized remains of an entire plant or animal are found. More likely, only hard parts such as tooth, bone and shell will be preserved. Even these won't be fossilized, if they are not buried quickly enough to prevent oxidation and decay. Once buried, the hard parts are often replaced by silica, carbonate or iron compounds in such a manner that details may, or may not, be preserved. Sometimes only molds or casts remain after hard parts dissolve. Fossils may also be destroyed by solution or recrystallization. Tracks and burrows are also considered to be fossils, since they too are records of life.

From this collection of bits and pieces, paleontologists assemble our knowledge of ancient life. Most of this information comes from Paleozoic and younger rocks, since an abundant and complex fossil record doesn't appear until the Cambrian. However, simple forms of life are found in rocks as old as 3 billion years. From these early beginnings, life evolved to become more complicated, diversified, and abundant. Not all groups of organisms flourished through time. Those which couldn't adapt to changes in their environments became extinct, while new groups appeared to take their places. From the geologic record we see that man is one of the relative newcomers to the long history of life on earth.

In addition to information on the evolution of life, fossils enable us to age-date and correlate sedimentary rock units as described in Chapter 1. They also provide us with a tool for reconstructing ancient environments. Using the law of uniformitarianism, we draw a comparison between the fossil record and the life requirements of similar modern plants or animals. Thus we come to think of solid, unfeeling rock as one part of a former living environment. It is this approach that enabled us to make statements about depth and temperature of the Silurian seas when we discussed reefs in Chapter 6.

Our region contains a fair diversity of fossils. The most common are Paleozoic marine invertebrate animals, which are quite abundant in some areas. Few plant and vertebrate fossils are known from this area, and most of these are found in Pleistocene deposits.

Invertebrates

The types of invertebrate fossils we find in our Paleozoic rocks indicate that the Cambrian through Devonian seas were generally warm and shallow. In this marine environment, microscopic and larger plants utilized the sun's energy and the dissolved mineral matter from the eroding land. A great variety of animals fed on these plants and each other. Life thrived, reproduced, died, and accumulated along with the sediments on the sea bottom. Some of these forms were ultimately fossilized. Locally, conditions were more favorable for preservation, so that some rocks are composed almost entirely of fossils (Fig. 7-3).

Representatives of all the invertebrate phyla and many of the classes are present in our Paleozoic rocks. Protozoans are single-celled animals whose presence can be detected with laboratory separation techniques and microscopic examination. Such efforts establish that foraminifera are common in the Devonian rocks of Wisconsin.

Figure 7-3. Cephalon and pygidium casts of the trilobite *Bumastus* from the Silurian Racine Dolomite; Hamilton, Illinois. The large cephalon with the well-preserved eye structure in the upper left corner of the sample is 2 inches across. The segmented fossil in the right center of the sample is part of a large crinoid stem. Sample courtesy of The University of Wisconsin-Milwaukee Greene Museum; photo by Frank C. Charnon.

A few sponges are known from this area, but recent work suggests that the distinctive Ordovician spongelike fossil, *Receptaculites,* is actually a colonial algae (Fig. 7-4). This form is sometimes called the sunflower coral because of its shape and the symmetrical pattern of the porelike openings.

The coelenterates include jellyfish, sea fans, and corals, but only the corals abundant fossils. Solitary horn-shaped corals, with well-developed vertical septa that divide the living chamber, are found in our Ordovician through Devonian rocks. *Streptelasma* is a familiar Ordovician example (Fig. 7-4). Colonial corals, with horizontally divided living chambers (tabulate corals), are

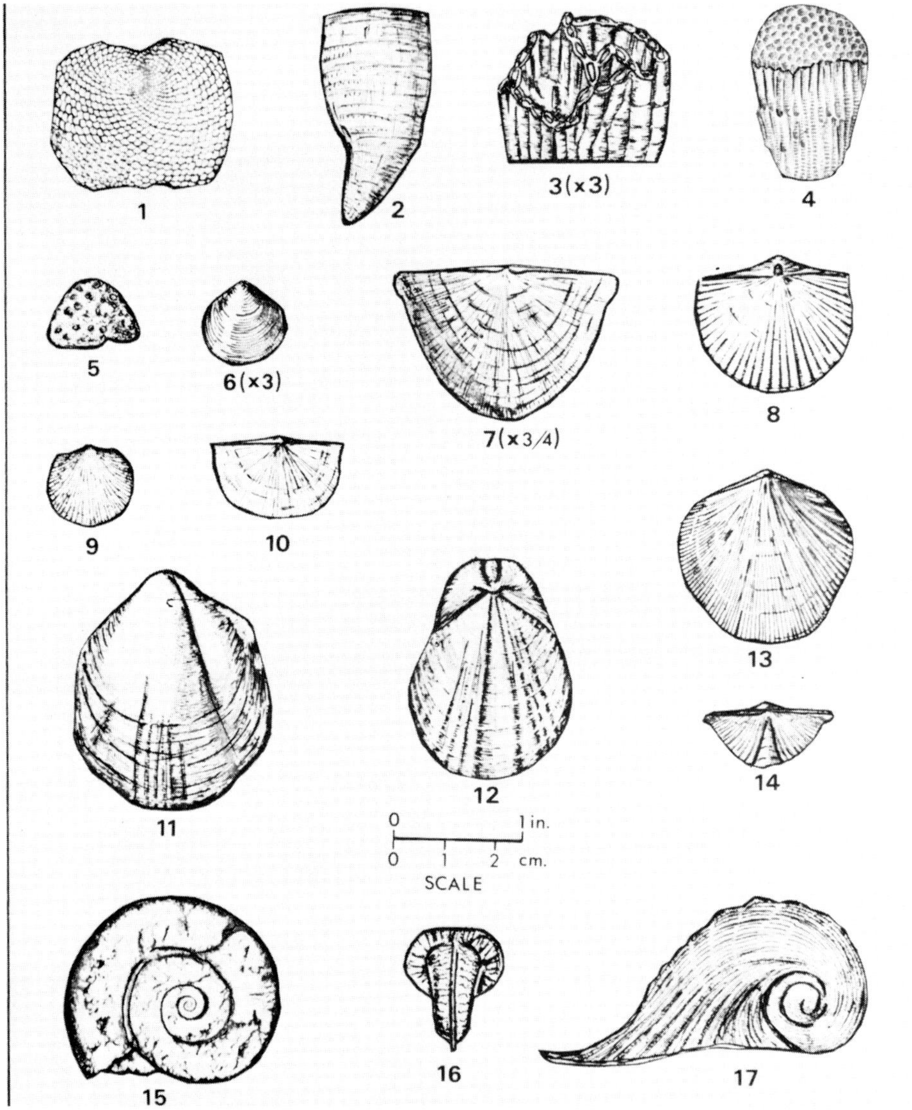

Figure 7-4. Representative Paleozoic fossils. All forms are to scale unless otherwise noted. (1) Top view of part of a *Receptaculites* colony—Ordovician, (2) Side view of *Streptelasma*—Ordovician, (3) *Halysites* (enlarged 3X)—Silurian, (4) *Favosites*—Silurian, (5) Side view of *Prasopora* colony—Ordovician, (6) *Dicellomus* (enlarged 3X)—Cambrian, (7) *Rafinesquina* (reduced 1/4)—Ordovician, (8) *Hesperorthis*—Ordovician, (9) *Resserella*—Ordovician, (10) *Sowerbyella*—Ordovician, (11) *Pentamerus*—Silurian, (12) *Virgiana*—Silurian, (13) *Atrypa*—Devonian, (14) *Mucrospirifer* (reduced 1/3)—Devonian, (15) *Maclurites*—Ordovician, (16) Back view of *Phragmolithes*—Ordovician, and (17) Side view of *Tremanotus*—Silurian.

especially common reef-builders in the Silurian. In fact, corals are such an important part of the Silurian fossil record that it is called the age of corals. The best known colonial corals are the chain coral *Halysites* and the honeycomb coral *Favosites* (Fig. 7-4). The Silurian reef-building stromatoporoids are usually considered to be lime-secreting relatives of the corals, but recently it was suggested that they may be sponges instead.

Such uncertainties regarding the affinities of fossils are a fundamental problem of paleontology. The biological definition of similar species requires that the individuals are capable of producing offspring, and this test isn't any help in defining extinct fossil species. Furthermore, paleontologists, unlike biologists, can't observe the relationship between an organism's hard and soft parts, nor can they take the creature home to determine its life-style.

Bryozoans of several distinctive types are common in our Ordovician through Devonian rocks. These animals, like the colonial corals, live in tiny, cuplike chambers as part of a colony that develops branching, fanlike, or lacy composite skeletons. Many of these resemble plants, hence their common name of "moss animals." Some occur as encrustations on other fossils while *Prasopora,* a common Ordovician variety, develops the configuration of a large gum drop (Fig. 7-4).

Brachiopods are bivalves somewhat like present-day clams or scallops, except that the opposing valves are different. Consequently, the only plane of symmetry in the shells of these animals is a longitudinal one through the beak. The simplest type of brachiopod is one that does not have an extended hinge line. Such types are essentially round (orbiculoids), conical (neotremes) or elongate (linguloids). These are the inarticulate brachiopods and they usually have shells of black, bluish or white phosphatic material, or a leatherlike exterior called chitin. Such fossils are common in the Cambrian of our region, and one of these genera is *Dicellomus* (Fig. 7-4). Articulate brachiopods generally have a well-defined hinge line and calcareous shells. This class is abundant in Ordovician and younger Paleozoic rocks. Some common Ordovician genera are: *Rafinesquina, Hesperorthis, Resserella,* and *Sowerbyella* (Fig. 7-4). In the Silurian, the pentamerid family, which includes *Pentamerus* and *Virgiana,* is especially distinctive (Fig. 7-4). The Devonian has an abundant brachiopod fauna that includes *Atrypa* and *Mucrospirifer* (Fig. 7-4).

The mollusks are a diverse phylum that includes the gastropods (snails), nautiloid cephalopods, and pelecypods (clams). Fossil gastropods can be coiled flat or into a high spire. In this region they are common in Ordovician through Devonian rocks. *Maclurites* is a flatly coiled Ordovician variety, while *Phragmolithes* and *Tremanotus* are genera with enlarged apertures (Fig. 7-4). Freshwater snails, much like those living in this region today, are present in Pleistocene lake deposits.

Nautiloid cephalopods are present in Ordovician through Devonian rocks. Their elongate, sutured shells are straight or gently curved, and close to a point at one end. The sutures, also straight or slightly curved, are formed where the septa meet the outer shell. Septa are the bases of a succession of former living chambers. A central canal extends through the septa, but most of the living animal occupied the largest chamber just within the open aperture. Common straight-shelled genera are *Michelinoceras* of the Ordovician and *Dawsonoceras* of the Silurian (Fig. 7-5). A curved form in Silurian rock is *Phragmoceras* (Fig. 7-5). These cephalopods had large eyes and numerous tentacles to gather food. They moved backwards by jetting water forward. Most genera were less than 3 feet long, but a giant form grew to 15 feet. Imagine this beast, with eyes flashing and tentacles waving, jetting through the warm, shallow Silurian sea toward you. This sort of thing would have hindered the development of Paleozoic oceanside resorts.

Pelecypods, or clams, are bivalved mollusks. Unlike the brachiopods, however, their only plane of symmetry is between the two valves. Thus, each valve is a mirror image of the other. Clams are not abundant fossils in this region, but they are present in Ordovician through Devonian rocks. Freshwater forms are also found in Pleistocene lake deposits.

The arthropod phylum includes trilobites, crustaceans, merostomes and insects. The trilobites, extinct relatives of the horseshoe crab, are the most important fossils. The name describes the three (tri) lobed subdivision of the chitinous shell of this animal into a broad head *(cephalon),* a central segmented body area *(thorax),* and a tail *(pygidium).* The shell of the trilobite was periodically shed like that of the modern crayfish and these molted shells helped enrich our record of these animals. Trilobites are present in all our Paleozoic rocks, but they are so abundant and useful for dating Cambrian rocks that this period is called the age of trilobites. Two Upper Cambrian trilobites found in our area are the small *Saukia,* and the large *Dikelocephalus* with its distinctive pygidium (Fig. 7-5). Other common types are *Isotellus* and *Ceraurus* from the Ordovician, *Bumastus* and *Calymene* from the Silurian, and *Phacops* from the Devonian (Fig. 7-5).

The other classes of arthropods are less common as fossils. Crustaceans are represented in our rocks by bivalved ostracods, that closely resemble small navy beans. A common form is *Leperditia* from the Ordovician (Fig. 7-5). Merostomes are somewhat similar in appearance to trilobites but they are biologically more complex. In our area they are occasionally found in the Lodi Shale Member of the St. Lawrence Formation. The insects are the most common animal in the world, but they are rarely fossilized. They are limited here to a few wood-boring beetles found within Pleistocene logs.

The echinoderms are marine animals that include starfish, sea cucumbers, sea urchins, blastoids, crinoids, and cystoids. In this area, only crinoids and cystoids are common fossils. They generally have oval heads (or *calyxes*) formed of calcareous plates, supported by stems of round carbonate disks which were attached to the sea floor. Food gathering arms are absent or poorly developed on cystoids but they are prominent on crinoids. Crinoids are often called sea lilies because they resemble flowers on a stem. Plates or parts of stems are common forms of fossil debris, but the fragile calyxes are rarely preserved. *Holocystites* is a cystoid and *Eucalyptocrinites* is a crinoid from the Silurian of this region (Fig. 7-5). Other genera are occasionally found in Ordovician and Devonian rocks.

Graptolites are extinct colonial animals represented by a narrow, straight or branched, chitinous skeleton. These are especially common in Ordovician and Silurian black shales, as carbon films preserved along bedding planes. Apparently what we find fossilized was a part of some soft-bodied floating or swimming organism. Graptolites are rare in our region, although specimens are locally present in the lower parts of the Ordovician Maquoketa Shale and the Silurian dolomite of eastern Iowa; an example is the Ordovician form, *Diplograptus* (Fig. 7-5).

A record of worms is also preserved in our Paleozoic rocks. One of these is the straight,

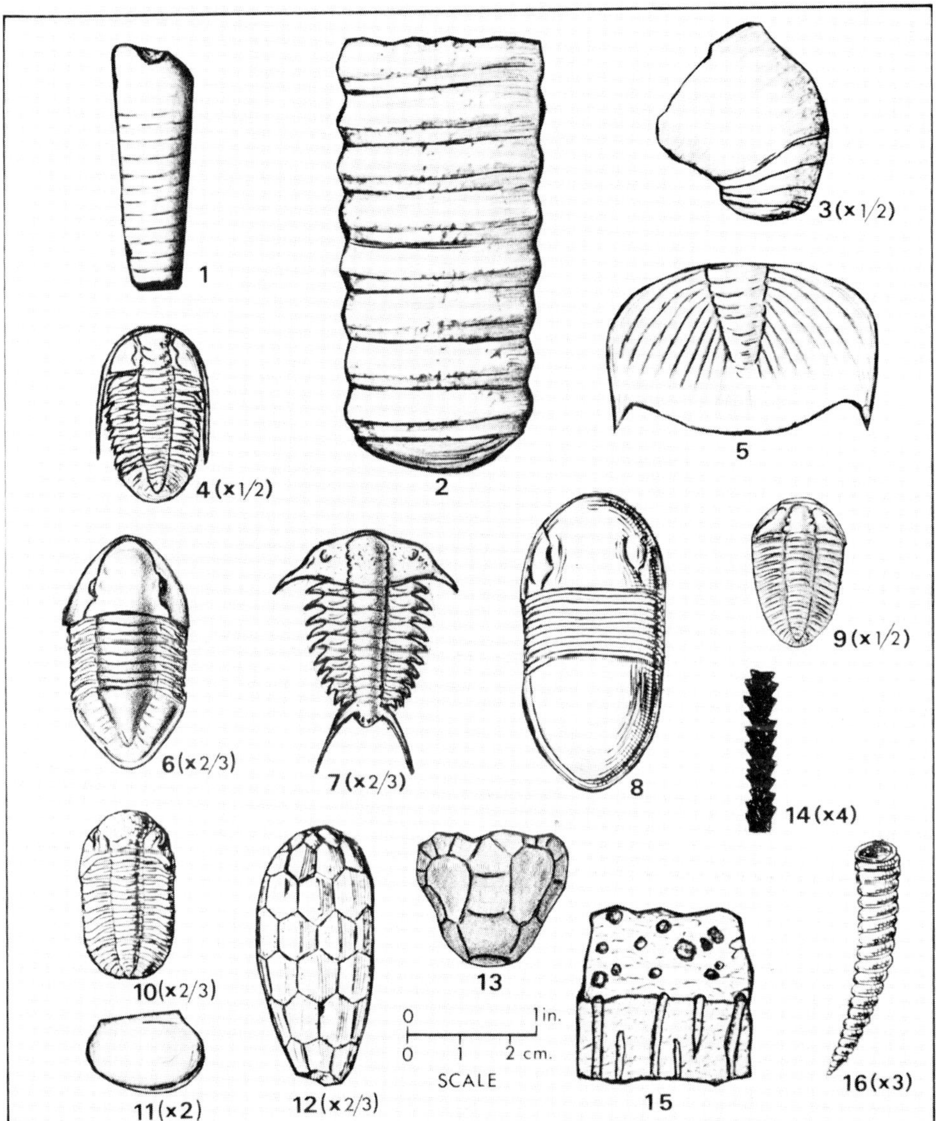

Figure 7-5. Additional representative Paleozoic fossils. All forms are to scale unless otherwise noted. (1) *Michelinoceras*—Ordovician, (2) *Dawsonoceras*—Silurian, (3) *Phragmoceras* (reduced 1/2)—Silurian, (4) *Saukia* (reduced 1/2)—Cambrian, (5) Pygidium of *Dikelocephalus*—Cambrian, (6) *Isotellus* (reduced 1/3)—Ordovician, (7) *Ceraurus* (reduced 1/3)—Ordovician, (8) *Bumastus*—Silurian, (9) *Calymene* (reduced 1/2)—Silurian, (10) *Phacops* (reduced 1/3)—Devonian, (11) *Leperditia* (enlarged 2X)—Ordovician, (12) *Holocystites* (reduced 1/3)—Silurian, (13) *Eucalyptocrinites*—Silurian, (14) *Diplograptus* (enlarged 4X)—Ordovician, (15) *Scolithus* borings in sandstone—Cambrian, (16) *Cornulites* (enlarged 3X)—Ordovician.

vertical boring of *Scolithus* that is common in Upper Cambrian sandstones (Fig. 7-5). Another is a small, tapered, corrugated tube called *Cornulites* that is locally abundant at the top of the Upper Ordovician Maquoketa (Fig. 7-5). Some carbonate rocks, when processed in a laboratory, will also yield worm jaws and conodonts. Although conodonts are microscopic phosphatic structures from an animal of unknown affinities, they are excellent aids in age-dating and correlation.

Plants and Vertebrates

Fossil plants and vertebrates are not abundant in this region, although a record of their presence extends from the Middle Precambrian to the Pleistocene. Our oldest fossils are microscopic algae and bacteria in Middle Precambrian rocks of northern Michigan. Large masses of colonial algae are also present in the Middle Precambrian Kona Dolomite, and small colonies are known from Upper Keweenawan rocks near Copper Harbor, Michigan (Fig. 2-8). Similar colonies are common in our Ordovician and Silurian rocks (Fig. 2-18).

Our oldest vertebrate fossils are the phosphatic fish plates found in the Middle Devonian Milwaukee Formation, and the fish teeth and bones from the Upper Devonian Kenwood (Antrim) Shale. The Kenwood also contains thin, coaly beds with poorly preserved plants. The general absence of deposition from Devonian to Pleistocene in our area limits the fossil record of this time interval to poorly preserved leaf impressions and some petrified wood from isolated patches of nonmarine Cretaceous sediments in the Driftless Area. It is unfortunate that our record of this long interval is so sparse, since during this time the great coal swamps flourished, dinosaurs roamed, flowering plants developed, and mammals appeared.

Our best plant and vertebrate fossils were buried in Pleistocene outwash and lake deposits. These include the Two Creeks forest bed described in Chapter 2, as well as other occurrences of birch, spruce, pine, and tamarack logs. Probably the most spectacular Pleistocene vertebrate fossil from our region is the mastodon skeleton found in a deposit of black clay near Richland Center, Wisconsin. A restoration of the skeleton of this huge beast is on display in the Geology Museum at the University of Wisconsin-Madison.

Teeth and tusks from mastodons and wooly mammoths are found in both the glaciated and nonglaciated parts of our region. These ancient elephants once roamed over North America, Europe and Asia, and the fossil record elsewhere shows that they were hunted by early man. They became extinct near the end of the Pleistocene, probably because of climatic changes, but some paleontologists suggest that early man was too successful as a hunter. Other Pleistocene vertebrate fossils from this region include an extinct peccary (a piglike animal), giant beaver, birds, fish, reindeer, and our long-enduring white-tailed deer.

Collecting Localities

The abundance and preservation of fossils in the Upper Cambrian is generally poor. The best fossils are brachiopods and trilobite fragments in the Lodi Member of the St. Lawrence, near the top of the Cambrian succession. The Lodi is quite poorly exposed, but outcrops include: cuts on Highway 44 near Kingston, Wisconsin (Green Lake County); cuts near the top of the hill on Highway 82 about 6 miles west of Mauston, Wisconsin (Juneau County); cuts along U.S. 14 between Cross Plains and Mazomanie, Wisconsin (Iowa County); quarries southeast of Richland Center, Wisconsin (Richland County); and the top of the bluff at Fire Bell Hill on Highway 182 just north of Lansing, Iowa (Allamakee County).

Other Upper Cambrian collecting localities for brachiopods, trilobites, and worm borings include: roadcut and adjacent hill at south edge of Hudson, Wisconsin (St. Croix County); quarry at top of Mt. Washington in Eau Claire, Wisconsin (Eau Claire County); cuts along Wisconsin 80 about 6 miles north of Elroy, Wisconsin (Juneau County); 75 feet up the bluff at Trempealeau, Wisconsin (Trempealeau County); cuts in the top of the Munising Formation along U.S. 2 about 0.5 miles north of Waucedah, Michigan (Dickinson County); on the north side of the old Pewabic mine near Iron Mountain, Michigan (Dickinson County); along U.S. 61 at the south edge of Lake City, Minnesota (Wabasha County); and along U.S. 61 at west edge of Red Wing, Minnesota (Goodhue County).

Fossils are not common or well preserved in the dolomite and sandstone of the Lower Ordovician Prairie du Chien Group. Those that are present are usually molds or casts of gastropods and corals. Colonial algae are locally abundant. A few fossils can be found wherever these rocks are exposed in road cuts or quarries along their outcrop belt in northern Michigan, Wisconsin, Illinois, Iowa, and Minnesota, but no specific locality is especially productive.

Middle Ordovician rocks include the St. Peter Sandstone at the base, the Glenwood Formation in the middle, and the Platteville-Decorah-Galena sequence at the top. The lower two formations in this succession are essentially unfossiliferous, although the St. Peter contains some worm borings, and microscopic worm jaws and conodonts are present in the Glenwood. The Platteville Formation is quite fossiliferous, and it contains pelecypods, brachiopods, bryozoa, cephalopods, gastropods, corals, ostracods, and trilobites. Preservation and collecting are best where this formation is only partially dolomitized. Such areas are in the western part of the lead-zinc district and in northern Michigan.

The Decorah Formation is a 35 to 45 foot interval of thin-bedded limestone and dolomite with some interbedded greenish shale. The fauna is similar to that found in the Platteville, although Decorah fossils are more abundant and preservation is generally better. The small biscuit-shaped bryozoan *Prasopora* is found at the top of the formation (Fig. 7-4). The Galena Dolomite is not especially fossiliferous and preservation is usually poor. However, the distinctive *Receptaculites* is common in this formation (Fig. 7-4).

Although Middle Ordovician rocks are widespread in our area, fossil collecting is best where the limy phase of the Platteville is developed or where the Decorah is well exposed. These favorable conditions are generally found in northern Michigan, southwestern Wisconsin, northwestern Illinois, eastern Iowa, and southeastern Minnesota. Specific locations in Michigan include: Bony Falls and Chandler Falls on the Escanaba River (Delta County); road, railroad and quarry cuts in the vicinity of Trenary (Alger County); cuts at Cornell (Delta County); the Groos quarry 5 miles north of Escanaba (Delta County); along the Rapid River immediately east of the town of Rapid River (Delta County); and in the vicinity of Spalding and Wilson (Menominee County).

Some Wisconsin Middle Ordovician localities are: quarries along U.S. 151 about 4 miles northeast of Sun Prairie (Dane County), cuts along U.S. 61 about 3.5 miles northwest of Dickeyville (Grant County), which provide exposures from the Platteville through the Galena; cuts along U.S. 151

about 7 miles southwest of Platteville (Grant County), which provide exposures from St. Peter through Platteville; quarries near Calamine (Lafayette County), cuts along County P about 1.5 miles north of Pine Bluff (Dane County), and cuts on Wisconsin 69 about 3 miles north of New Glarus (Green Lake County).

Middle Ordovician collecting localities in Illinois include numerous quarries and road-cuts in Jo Daviess, Stephenson, and Carroll counties. In Iowa, some specific localities are: cuts in the Platteville Formation immediately west and south of McGregor (Clayton County), cuts in the Decorah Formation at junction of U.S. 52 and County C at Guttenberg (Clayton County), and cuts in the Galena Dolomite one mile south of Dubuque along U.S. 52, 67, and 151 (Dubuque County). Minnesota collecting localities include quarries and cuts along the road toward White Rock, just off U.S. 52 south of Cannon Falls (Goodhue County).

The Upper Ordovician Maquoketa Shale is richly fossiliferous throughout its outcrop belt, but it is rarely exposed. Where it is present, bryozoa and brachiopods are usually found in thin-bedded limestone interbedded with greenish shale. The distinctive Depauperate Zone is present at the base of the Maquoketa. This zone of abundant, very small pelecypods, gastropods, and cephalopods was once considered to contain a dwarf fauna that reflected a muddy, oxygen-deficient marine environment. Recent work indicates that most of the more than 40 species in this fauna are small forms and not dwarfs of larger ones. One of the best known Upper Ordovician fossil localities is near Graf, Iowa. Here, the base of the Depauperate Zone is overlain by a zone of uncrushed graptolites (*Diplograptus*) (Fig. 7-5). Above the graptolite zone is a bed that is solidly packed with straight-shelled cephalopods (*Isorthoceras*). Another interesting fossil occurrence is the *Cornulites* Zone that is sometimes present at the top of the Maquoketa in northwestern Illinois and eastern Iowa (Fig. 7-5).

Some fossiliferous outcroppings of the Maquoketa are: along the Little Bay de Noc shore of the Stonington Peninsula (Delta County), Michigan; at Wequiock Falls Roadside Park along Wisconsin 57 about 10 miles north of Green Bay (Brown County), Wisconsin; a quarry just south of Oakfield (Fond du Lac County), Wisconsin; and cuts into the Depauperate Zone at the junction of County O and W, about 2 miles south of Shullsburg (Lafayette County), Wisconsin.

In Illinois, Maquoketa outcrops include: a railroad cut into the Depauperate Zone near the village of Scales Mound (Jo Daviess County), a private quarry exposing the Ordovician-Silurian contact along U.S. 20 about 4 miles west of Marengo (McHenry County), and in the vicinity of Mt. Carroll in Carroll County. Some exposures in Iowa are: along Mill Creek in Bellevue State Park (Jackson County), road and railroad cuts immediately southwest of Graf (Dubuque County), and along the Little Maquoketa River northeast of Graf.

Silurian dolomites are sparsely to moderately fossiliferous throughout our area. Preservation in these dolomitized rocks is generally limited to casts and molds, although some of these reproduce many details of the original shell. In a few instances the fossils are replaced with silica. Common fossils include corals, stromatoporoids, trilobites, brachiopods, cephalopods, and crinoids. A representative collection of these fossils is available for study at the Greene Museum of the University of Wisconsin-Milwaukee.

The resistant dolomites of the Silurian are well-exposed at many places along the outcrop belt. Specific localities in Michigan are at Burnt Bluff near Fayette on the Garden

Peninsula (Delta County), and just east of our region in the old quarries around Manistique (Schoolcraft County).

In Wisconsin, exposures include: road-cuts, lake cliffs, numerous quarries and rock fence rows throughout the Door County Peninsula; abandoned quarries south of Oakfield (Fond du Lac County), quarries in the Cedarburg and Grafton area (Ozaukee County), quarries around Lannon (Waukesha County), and quarries north of Racine (Racine County).

The Silurian in Illinois is exposed in several places in the vicinity of Chicago (Cook County), in the Mt. Carroll area (Carroll County), and on the bluffs along the Mississippi River in northwestern Carroll County and southwestern Jo Daviess County.

Iowa exposures include: cuts along old U.S. 20 about 3 miles east of Centralia (Dubuque County), cuts along U.S. 52 about 8.5 miles south of Dubuque (Dubuque County). In Iowa's Bellevue State Park (Jackson County), a graptolite locality occurs 12 feet above the base of the Silurian dolomite.

The Devonian in our area is limited to eastern Wisconsin (Milwaukee, Ozaukee, and Sheboygan counties), and outcrops are rare. Parts of the Devonian yield a rich and varied fauna that includes brachiopods, cephalopods, gastropods, bryozoans, trilobites, crinoids, corals and fish plates. A collection of these fossils is on display at the Greene Museum at the University of Wisconsin-Milwaukee. Exposures include those along the Milwaukee River in Estabrook Park in the northeast part of Milwaukee, and at Harrington Beach State Park in Ozaukee County. Fossiliferous Devonian rocks are also exposed adjacent to our area in east-central Iowa, in north-central Illinois, and in the northeastern part of the Lower Peninsula of Michigan.

Chapter 8

GEOLOGIC RESOURCES

The geologic resources of a region include more than the mineral deposits and mineral industries. The agricultural soils, the supplies of surface and subsurface water, the scenic beauty, and many of the recreational and tourist attractions are also important geologic resources. In our area, the soils, rivers, lakes, and scenery are products of a long and complex geologic history, but Pleistocene glaciation provides the strongest imprint.

Mineral production from this region is dominated by iron ore and nonmetallic minerals such as sand and gravel. Smaller amounts of copper, lead and zinc are produced, but most of the mineral products required by our industries are mined elsewhere. In Wisconsin, sand and gravel, and stone products contribute about 40% and 30% respectively to the value of the state's mineral production. Metallic ores of iron, zinc, and lead provide another 20%, while lime, clay, peat, and miscellaneous mineral products make up the remainder.

The value of mineral production in northern Michigan is concentrated in iron, with more than 16 million tons of pelletized ore shipped in 1974. Michigan copper production is currently limited to the White Pine mine in Ontonagon County, but economics could curtail this operation. A variety of nonmetallic minerals are also produced in the Upper Peninsula.

In northern Illinois, sand, gravel, and stone dominate production, with lead and zinc contributing less than 3%. No metallic minerals are mined in southeastern Minnesota and eastern Iowa, but nonmetallic minerals are produced.

There is no production of precious metals or gems from this region, although a gold mine operated north of Ishpeming, Michigan from 1883-1897. With high gold prices, there is renewed interest in this property and it may become productive once again. The discovery of low-grade silver ore at Silver City in Ontonagon County, Michigan caused a brief flurry of exploration around 1860, but no mining developed. However, significant amounts of silver are recovered along with copper at the White Pine mine, south of Silver City. About twenty diamonds were found in the glacial drift of Wisconsin and this provided the inspiration for the diamond swindle previously described. Unfortunately, no diamonds were found after 1885.

Coal and oil are not produced in this area. The existing sedimentary rocks were deposited prior to the time of widespread coal deposition and geologic conditions are not favorable for petroleum occurrences. However, commercial petroleum production in adjacent areas of Iowa, Illinois, and Michigan has encouraged periodic petroleum exploration through the years.

Recent mineral exploration in the Northern Highland has resulted in renewed interest in the mineral potential of this province. Geophysical and geochemical surveys by major corporations and governmental agencies have stimulated the leasing of acreage throughout this area. Copper, zinc, nickel, uranium, gold, and silver mineralization associated with Precambrian igneous rocks are the objects of this search. With a favorable economy and improved technology, new ore bodies will be located for future use.

SAND AND GRAVEL

Our widespread, abundant sand and gravel deposits are largely a gift from the Pleistocene. The task of grinding, washing, and sorting bedrock fragments and minerals was performed by the lobes of glacial ice, and the floods of meltwater which issued from the ice margins. A visit to a gravel pit, or even a gravel road or driveway, provides a sampler of the bedrock along the route of the ice advance. The knowledgeable observer will find various rocks and minerals from northern outcrop areas mixed with more common types from local sources.

The areas of greatest sand and gravel production are located in the urbanized parts of southeastern Wisconsin and northeastern Illinois, where deposits near the now vanished ice margins are abundant. This fortunate circumstance helps reduce the cost of road building and construction in this region.

CRUSHED STONE

In areas where Pleistocene gravel supplies are not plentiful, local bedrock is used for construction aggregate or road ballast. Dolomite, limestone, quartzite, and less frequently granite and basalt, are crushed for these purposes. In a few instances, highly weathered granite is used. Industries requiring rock fragments with particular properties of hardness, durability or color, utilize rhyolite and basalt for roofing granules, dolomitic marble for terrazo tile, and quartzite for an abrasive. The Precambrian Baraboo Quartzite is crushed at Rock Springs, Wisconsin to provide railroad ballast. Crushed, friable, pure quartz sandstone is the source of silica sand products, such as those used by the foundry and glass industries. Finely crushed limestone and dolomite are still used as agricultural lime because of their lower cost, but chemical lime is preferred.

Figure 8-1. Partially restored lime kilns along the Niagaran cuesta at historic Lime Kiln Park, Menomonee Falls, Wisconsin.

LIME AND CEMENT

To produce lime, limestone or dolomite is crushed and then heated to drive off carbon dioxide. This reduces the volume by about one-half. Since lime releases heat when it combines with moisture, a controlled amount of water is frequently added to the finished product to avoid danger of fire. The resulting hydrated lime is the primary commercial lime product.

This area was one of the leading lime producers in the United States at the turn of the century, and old lime kilns are still present today in many communities (Fig. 8-1). Silurian dolomite is an excellent raw material for the lime industry, and this is reflected by the concentration of lime plants, past and present, along its outcrop belt where suitable transportation routes are also available. The initial uses for lime were primarily in the agricultural and construction industries. Today, lime is also used as a chemical in a variety of industries, including those which produce metals, paper, and glass. It is also used for water purification and in waste treatment plants.

The carbonate bedrock in this region generally has too high a magnesium content to be suitable for the manufacture of Portland cement, and for many of the uses of chemical lime. However, one cement plant at Dixon, Illinois (Lee County) has used Ordovician limestone for many years, and several large quarries in the Upper Peninsula, east of our area of concern, provide carbonate for a flux in steel furnaces.

The Devonian Milwaukee Formation of Wisconsin was utilized as a natural cement rock prior to the turn of the century, and the old quarry tunnels underlie parts of metropolitan Milwaukee. Cement plants still operate in Wisconsin at Milwaukee and at Manitowoc, but the limestone they use is imported.

BUILDING STONE

Dimension or building stone of several varieties was once quarried throughout the area. Increasing costs of quarrying and building with stone caused a decline in this industry. In the middle of the last century in the lead-zinc district, it was much cheaper to build with the local carbonate bedrock rather than wood. Consequently, the poor built many of the stone houses that are now so prized in this region. In Wisconsin, the Silurian dolomite is quarried as a building stone in a number of places, but especially near Lannon (Waukesha County) (Fig. 8-2). This rock, called Lannon Stone, has excellent qualities for building. Sandstone is still cut at a few places in Wisconsin, but the deep red or brown sandstones along the shores of Lake Superior in Wisconsin and northern Michigan, that were highly prized for construction during the early part of this century, are no longer quarried.

High quality igneous and metamorphic rock for building and landscaping is also available in Wisconsin and northern Michigan. Rock types, with an infinite variety of color and texture, include granite, rhyolite, gabbro, schist, slate, marble, amphibolite, and quartzite. One active quarrying area is about 10 miles north of Wausau in Marathon County, Wisconsin. A distinctive red granite, the state rock of Wisconsin, is cut

Figure 8-2. Well-bedded Silurian dolomite in an abandoned quarry at High Cliff State Park, Calumet County, Wisconsin. The flat floor of this quarry illustrates the even, persistent nature of the beds in this formation. Wisconsin Department of Natural Resources photo.

here. In the past, Wisconsin quarry sites included: Montello (granite), Waupaca (granite), Berlin (rhyolite), and Mellen (gabbro), to name a few. In Michigan's Dickinson County, amphibolite and feldspar-rich rocks are quarried near Randville, while marble is quarried at Felch. Igneous and metamorphic rocks are generally more costly to produce than sedimentary rocks, and consequently they are usually reserved for monuments or decorative stonework.

CLAY AND SHALE

Clay and shale are quarried for the brick and tile industries and for an additive to cement. The clay-rich Ordovician Maquoketa Shale is mined near Oakfield in Fond du Lac County, Wisconsin, and commercial concentrations of Pleistocene clay are exploited in Racine County, Wisconsin and in the Chicago area. Clay deposits of both types are scattered throughout our area, but their development depends on transportation, firing cost and market demand. Since the use of brick for building purposes declined, clay production also decreased, and most of the local brickyards and tile factories are closed. The Red Wing Pottery works at Red Wing, Minnesota (Goodhue County) used Cretaceous clay from a pit near Goodhue for its products until a few years ago when mining ceased. Clay products are still sold by this firm, but they are now made elsewhere.

PEAT

This region has large reserves of peat which is used as a seed inoculant and as a general soil conditioner. In Wisconsin, most of the peat produced comes from Waukesha County, while Illinois production is concentrated in the greater Chicago area. None is mined in upper Michigan.

Peat was used as a fuel around the turn of the century since it was a familiar energy source to early settlers from northern Europe. The largest and thickest peat deposits are in bogs and old lake beds of the glaciated area, and in the area once occupied by Glacial Lake Wisconsin.

LEAD AND ZINC

Lead and zinc deposits have been known in the Upper Mississippi Valley region since occurrences were reported by French explorers and missionaries in the late 17th century. Nicholas Perrot is generally credited with recognition of these deposits, and he was probably the first exploiter of the ores. In 1697, Pierre Le Sueur received permission to work the ores near what is now Galena, Illinois and in other mineralized areas along the Mississippi River. He lacked geological knowledge, however, and shipped 3,000 pounds of green, glauconitic Cambrian sand (glauconite is a complex iron and magnesium silicate mineral) home to France, assuming it was copper ore. Imagine his embarrassment after contemplating the long canoe and ship journey of this worthless cargo!

Lead mining was the earliest mineral industry in the upper Midwest. From a modest beginning, it expanded into a thriving business by the mid-1800s before the rich deposits played out. Tributes to the primary lead mineral, which created the boom in the Driftless Area, occurred in 1827 when La-Point, Illinois was renamed Galena, in 1880 when New Galena, Iowa was founded, and most recently when galena was named the state mineral of Wisconsin.

The lead-zinc ore is found in Ordovician carbonates where sulfides were deposited by mineral-bearing waters of moderate temperature at shallow depth. Patterns of mineralization relate to systems of minor folds, faults, and joints, but they also reflect fluid migration along bedding planes and into cavities.

Zinc minerals, primarily sphalerite (zinc sulfide), were originally separated from the lead ore and discarded. It wasn't until 1860 that a satisfactory smelting method created a market for zinc. Zinc soon surpassed lead in economic importance, and today the combined production of lead and zinc is tied to the price of zinc, which constitutes the major metal in the ore. The original productive area is called the Upper Mississippi Valley Lead-Zinc District, and it included northwestern Illinois, southwestern Wisconsin, and northeastern Iowa. Currently, high operating costs and poor market conditions have closed almost all of the mines in the district except those operated by Eagle-Picher Industries near Shullsberg, Wisconsin and Galena, Illinois. The ore from these mines is processed at Shullsberg and at the Graham mill north of Galena.

COPPER

Copper mining in our region is restricted at present to the low-grade copper sulfide (chalcocite) ores being extracted at White Pine, in Ontonagon County, Michigan, by White Pine Mining Company (a subsidiary of Copper Range Company). The chalcocite, with a small amount of native copper and lesser amounts of silver, is disseminated in the Nonesuch Shale of Late Keweenawan age. Averaging less that 2% copper, the ore comes from four narrow shale and siltstone zones in the lowermost 20-25 feet of the formation. The great size of the deposit makes profitable production possible, and the result is one of the largest underground copper mines in the United States.

The Nonsuch sediments accumulated on a vast delta which spread into an ancient sea from a highland area to the south. The copper-bearing horizons contain varying quantities of finely dispersed carbonaceous material, and the richer copper-bearing zones are those with the most organic material. Petroleum residues also occur in pore spaces in these one billion year old units. It is not known if the copper minerals were deposited during sedimentation, or if they were replacement deposits formed by ore-bearing waters of moderate temperature. However, it seems that carbonaceous material influenced the distribution of copper in some manner.

Forty-five miles northeast of the White Pine mine is the famous native copper mining area of the Keweenaw Peninsula. Here, copper is found in vesicles or cracks within Keweenawan basaltic lava or in the voids of porous, interbedded conglomerate. No longer mined, these deposits were exploited continuously, from soon after the publication of Douglass Houghton's famous geologic report on this area in 1841, up to 1968. Prospectors flocked to the Keweenaw during the early years of the mining boom and production accelerated. When the rapidly expanding electrical industry needed great quantities of copper in the early 1900s, it was largely supplied by these mines. Eventually, the most accessible deposits were mined out and the deep underground mines could no longer compete with western United States copper sources. Production ceased in 1968 but reevaluation of this historic district is now underway. Several mines (Centennial and Quincy) are being opened for exploration and an experimental mill is operating. Hopefully, economic ore will be found. More information on Keweenaw copper was previously provided in the special features section of Chapter 3.

Small occurrences of copper are present in the Keweenawan rocks of northwestern Wisconsin. The Douglas, St. Croix, and Minong

Copper Ranges trend across an upland between the Douglas and Lake Owen faults. This upland is called the St. Croix horst and it occupies parts of Douglas, Bayfield, Washburn, Burnett, Polk and adjacent counties. More resistant basalts create the ridge crests and backslopes, while less resistant layers form lowlands between the ridges. Copper occurrences are scattered along these ranges, but no commercial mining developed.

As the value of copper increased, exploration in Wisconsin became active once again. The Flambeau Mining Corporation, a subsidiary of Kennecott Copper Company, expects to begin an open pit mining operation near Ladysmith in Rusk County, in the near future. EXXON USA announced the discovery of a major copper-zinc sulfide deposit in 1976, which is larger than the Rusk County ore body. It is located in Forest County, 6 miles south of Crandon. Other prospects in Marathon County, east of Rhinelander in Oneida County, and in the Pembine-Amberg area of Marinette County are being studied.

IRON

Iron forms the kingpin of the mining industry in the Upper Peninsula of Michigan. It is currently extracted by Cleveland Cliffs from four mines in the Marquette Range (Tilden, Empire, Republic, and Mather), and from two locations in the Menominee Range; at Iron River (Inland Steel's Sherwood mine), and north of Iron Mountain (Hanna Mining's Groveland mine). Four of these six active mining sites are open pit operations while the Mather and Sherwood are underground mines. The Mather mine at Ishpeming is now taking ore from 3,500 feet below the surface, while the Sherwood at Iron River is working 1,875 feet underground.

Inland's Sherwood mine is the only operation that is shipping natural ore. The others utilize types of taconite iron formation that are upgraded by grinding, flotation and magnetic separation. After a final pelletizing operation, the product contains about 60% iron.

The importance of technology upon the economics of Upper Peninsula iron mining is well illustrated by the successful operation of Cleveland Cliff's Tilden mine northwest of Marquette. Here, a specialized flotation process was devised to extract finely disseminated, nonmagnetic iron particles that previously couldn't be separated. This mine has reserves of low-grade martite-hematite ore which are expected to last for well over a century.

During the period 1880 to 1932, iron ore was produced in the Wisconsin part of the Menominee Range in the Florence District. The Penokee-Gogebic Range of Wisconsin and Michigan was also once an active iron mining region. Hematitic ores were extensively mined in the Ironwood, Michigan and Hurley, Wisconsin areas from about 1885 to 1967 when the cost of following the steeply dipping ores underground became prohibitive (Fig. 8-3). Additional details on the iron ranges of Michigan and Wisconsin were previously described in the special features section of Chapter 3.

Iron ore was also produced from three other Wisconsin districts in the past. A small deposit was mined at Ironton, Wisconsin from about 1850 to 1875. Relatively large-scale operations began in the mid-1800s with production of the soft, oolitic hematite ore of the Ordovician Neda Formation along the Silurian escarpment in southeastern Wisconsin (Dodge County). This iron formation averages 15 feet in thickness, and it is overlain by about 20 feet of Silurian dolomite. Mining was initially underground, but later an open pit was put into operation. The ore was

Figure 8-3. Ore cars being loaded underground in the Penokee-Gogebic Range, about 1939; Montreal mine near Hurley, Wisconsin. Milwaukee Journal photo.

smelted locally, first at Neda and later at Mayville. Operations ceased in 1928 because these ores are relatively high in sulfur and phosphorus, which makes them undesirable for modern steel production.

The Baraboo Range was the focus of another iron mining industry in the first quarter of this century. Ore production ended when complex geological structure and groundwater flooding problems increased the cost of operations beyond economical levels.

At present, Wisconsin's only iron mine is in the Black River Falls District of Jackson County, where the production of ore began in 1969 (Fig. 8-4). At this mine, the Jackson County Iron Company (a subsidiary of Inland Steel) beneficiates low-grade magnetic ore into pellets for rail shipments to mills in East Chicago, Indiana. Ore reserves at the Jackson mine are estimated at 15 years with the present rate of production. Also of interest here is the mine development, which includes a plan for land reclamation after the ore is depleted. Tree and grass plantings on waste piles began in 1971, and the mine serves as a model of environmental concern for the mining industry.

Another iron mining district is located just west of our region in southeastern Minnesota. In this area, mining activity during 1942-1968 was concentrated in western Fillmore County, southeast of Spring Valley, Minnesota. The ore was generally the yellowish-brown oxide of iron called limonite. These deposits resulted from the deep

Figure 8-4. Multilevel open pit taconite mine and processing plant of Jackson County Iron Company, Black River Falls, Wisconsin. Milwaukee Journal photo.

weathering of Paleozoic carbonate bedrock. Testimony to the extent and effectiveness of weathering in this area is also provided by numerous sinkholes and caves. In spite of the closing of this district, iron mining still flourishes in the historic iron ranges of northeastern Minnesota.

Northeastern Iowa had an operating iron mine at Iron Hill near Waukon (Allamakee County) from 1913 to 1918. The ore body, like those in southeastern Minnesota, was formed from weathered Paleozoic carbonates.

POTENTIAL RESOURCES

The current surge of prospecting and exploration throughout the Northern Highland may result in the discovery of additional mineable copper, zinc, and iron deposits. Uranium prospecting is also underway in northern Waupaca County, Wisconsin and adjacent areas, and between Marquette and Iron River, Michigan. This activity reflects our growing dependance on atomic energy.

Other minerals are known to occur in some abundance within the Northern Highland but the deposits are not developed because of inadequate reserves or poor quality. Among these are asbestos, beryl, feldspar, kyanite, molybdenite, nepheline, talc, thorium, and zircon.

Petroleum exploration in eastern Wisconsin, eastern Iowa, and northern Illinois is periodically attempted by the wildly optimistic. At some future time, offshore production in Lake Michigan may be possible, but expectations for the rest of this region are very poor.

In spite of this limited potential, there are occasional hoaxes involving petroleum within this area. An early one took place in Monroe County, Wisconsin in 1865 when a gentleman from LaCrosse purchased land near Sparta. This land was not bountiful enough to suit him, so by night he planted several barrels of crude oil connected to a system of underground pipes and valves. Soon oil began to flow from the ground and he generously permitted friends and townspeople to buy stock in his Gem Petroleum Company. Amazingly enough, the flowage always improved when prospective stockholders were around. One day it "failed" and he left town $50,000 richer.

Coincidentally, a mineral water spring with an initial flow of 100 gallons a minute was later struck at the site of one of the oil "flows." A spa was developed to take advantage of these waters, which then were considered to be helpful for a variety of ailments.

EXTRACTION METHODS

Both underground and open pit methods are used to mine mineral resources. The geometry and size of the mineral occurrence, its depth, enclosing rock type, the amount of overburden, and environmental concerns determine the type of extraction that is utilized.

The lead-zinc deposits of southwestern Wisconsin and northwestern Illinois are mined underground. Underground mining for copper and iron continues in the Upper Peninsula, but most of the iron ores are now taken from open pit mines. In Wisconsin, the Black River Falls iron mine and the Ladysmith copper deposit (Rusk County) are carefully planned open pit operations. All of the nonmetallics (gravel, silica sand, clay, stone, and peat) are produced from open pits or surface quarries.

Land disturbed by surface mineral extraction in areas where land values are low may be left unavailable for any other use after mining activity has terminated. Out-of-sight is not out-of-mind, and underground mining also has its problems. Land subsidence and caving, improperly closed shafts, abandoned machinery and surface structures may result in hazardous nuisances. Excavations, either dry or water-filled, tailings ponds, and waste rock piles are often dangerous and unsightly. Such features are also subject to erosion and may contribute to water and air pollution.

The U.S. Bureau of Mines reports that 0.13% of Wisconsin's total land area was disturbed by mining activities between 1930 and 1971. This percentage would be similar for eastern Iowa and northern Illinois, and somewhat higher for Upper Michigan. About 25% of the disturbed acreage was reclaimed or restored to a usable state by the mineral industries, while additional acreage was restored by private individuals, developers, the state, or by communities.

As mineral exploitation increases to meet market demands, provision for sequential land use should be part of the planning stage of any mining project. At present, legislation concerning reclamation is limited and locally variable. For example, in Wisconsin, only the metallic mineral industries, which constitute about 2% of the state's total mineral extraction, are subject to stringent regulations. Products such as sand, gravel, and stone are generally extracted near a local market because of their high weight and relatively low value. Planning for the impact of such production on post-extraction land use is especially important in an urban area. Planning in another sense is also necessary, or many valuable deposits will be "urbanized'' before extraction, thus resulting in the permanent loss of the resource.

SOILS

The glaciers that covered much of this region during the Pleistocene influenced the supply and distribution of sand, gravel, clay, and peat deposits. This same Pleistocene heritage had a profound effect on soils, surface water distribution, and scenic resources.

Soils are our richest geologic resource, and their differences result from the nature of the bedrock and the age, thickness, and composition of the glacial drift (Fig. 8-5). Therefore, it is not surprising that soil maps are somewhat similar to the bedrock and the glacial maps, and consequently resemble the physical provinces map of this region (Fig. 3-1).

Figure 8-5. Soil profile developed on glacial outwash sediments near West Bend, Wisconsin. The dark, upper layer is organically enriched by decaying vegetation. The lighter layer beside the shovel blade is outwash sand and gravel.

The soil fertility of much of this region is derived from the glacially ground rock and mineral debris deposited during the Pleistocene. Only in the unglaciated areas have residual soils developed directly upon bedrock. Even here, a Pleistocene blanket of wind-deposited loess provides a bonus of fertility.

The quality of the drainage, degree of slope, and local differences in climate and native vegetation also influence soil development. These factors, when combined with bedrock and glacial variations, result in more than 350 distinctly different soils in Wisconsin alone. These soils have differing capabilities and thus impose limitations on land use. They determine the quality of agriculture and forestry in our region, and are important factors in planning and construction because of their differing drainage properties, strengths, and stabilities. The ideal concept in soil science is to utilize a soil with full knowledge of its best assets, and with due provision for its liabilities. The details are complex, but in general, soils are best in the southern part of our region, while those that are less desirable are in the central and northern areas.

WATER

This region has an abundant fresh water supply. The key to this renewable wealth lies in an annual rainfall averaging more than 30 inches per year, coupled with an evaporation rate which permits the storage of large quantities of both surface and subsurface water.

Let's consider our geographic setting.

Two of the Great Lakes form large parts of our borders, and the Mississippi, the master river of the United States, flows along the western edge of our area. About 10,000 inland lakes dot the glaciated regions, with significant concentrations in northern Wisconsin. It seems safe to say that our surface water resources are adequate, if we treat them with respect.

Ground water is an integral part of the entire water system, and as described in Chapter 1, it is directly related to surface water. This resource represents about 6 times that available from all surface sources. Its availability depends not only on recharge by precipitation and infiltration, but also on the character and thickness of the glacial drift and the bedrock beneath. Both can serve as aquifers when they have sufficient porosity for storage and the ability to allow the passage of water.

However, ground water supplies are not equally distributed over the region, since precipitation, runoff, infiltration, and aquifer characteristics are variable. Subsurface water is generally available in greatest quantity from the Cambrian sandstone aquifer and the Silurian dolomite. Thick glacial outwash deposits and deep valley alluvium also serve as high capacity sources. In near-surface aquifers, the water supply will vary seasonally, with higher water levels in spring. The northern part of our region is underlain by impermeable, crystalline Precambrian bedrock with a thin cover of glacial till. This combination results in generally unfavorable ground water conditions.

Although only 20% of the water used in the United States is from ground water at present, this resource has advantages in its immediate proximity and its constancy of quality and temperature. These factors caused ground water use to increase 145% nationwide between 1940 and 1971. The greatest growth in usage in our region has been by industry, and for irrigated farming.

If the rate of ground water use exceeds the rate of recharge, the water table drops and the available supply decreases. Due to the relatively slow recharge of subsurface water, there is need for regulated use wherever this resource is being depleted. Ground water is also prone to chemical and biological pollution, and at present, a concern for quality protection is of greater importance than availability in our region.

SCENIC AND RECREATIONAL

The scenic and recreational resources of this region are partially determined by earlier geologic history, but particularly by Pleistocene glaciation, which influenced the topography, drainage, and soils. These in turn, with due consideration for climate, control the natural vegetation and wildlife. Thus, geology casts the straw in favor of tamarack over oak, bass over trout, bog over sand barren, and mallard over grouse. All the recreational and resort industries, whether they emphasize scenery, hunting, fishing, camping, skiing, boating, or swimming are directly dependent on this control.

Glacial erosion and deposition, coupled with the relatively minor landscape modifying processes since the end of the Pleistocene, are responsible for most of our lakes. Many of our scenic parks are also products of the Pleistocene. The units of the Ice Age Scientific Reserve, especially the Kettle Moraine Forest, Devils Lake and the St. Croix Dalles, dramatically illustrate the icy impact on the landscape. The Dells of the Wisconsin River, the great bed of Glacial Lake Wisconsin with its castellated buttes,

Figure 8-6. Pebble beach and the Niagaran escarpment along the Green Bay shoreline, Door County, Wisconsin. Wisconsin Department of Natural Resources photo.

and the parks along the shores of lakes Superior and Michigan are also monuments to this great modifier of our landscape (Fig. 8-6).

The natural beauty of this region is a great geologic resource that cannot be inventoried by tons of ore produced, bushels of grain per acre, dollars per ton of sand and gravel, or cords of pulp wood per forty acres. Nevertheless, this resource contributes substantially to the personal wealth of those who seek its riches.

Appendix

REFERENCES CITED AND ADDITIONAL SOURCES OF INFORMATION

There is abundant information that provides details on the geology of this region. We have used much of this material in compiling this book, but with the exception of specific comments in the introduction and on some of the figures, we have not referenced our sources as one would in a technical publication. These specific references are listed below. However, readers may wish additional detail on some areas, and consequently suggestions for additional sources of information are provided.

The Public Information Office of the U.S. Geological Survey, Reston, Virginia 22092, will provide a free listing of all the federal survey publications for a given state. In addition, state geological surveys provide free listings of available publications. The addresses of these surveys are:

(1) Illinois State Geological Survey
Urbana, Illinois 61801

(2) Iowa Geological Survey
16 W. Jefferson Street
Iowa City, Iowa 52242

(3) Department of Natural Resources
Michigan Geological Survey
Stevens T. Mason Building
Lansing, Michigan 48926

(4) Minnesota Geological Survey
University of Minnesota
1633 Eustis Street
St. Paul, Minnesota 55108

(5) Wisconsin Geological and Natural History Survey
University of Wisconsin-Extension
1815 University Avenue
Madison, Wisconsin 53706

Geologic and topographic maps are also important sources of information. An index map that details the published geologic maps for a given state may be purchased at a modest price from the U.S. Geologic Survey. The U.S. Geological Survey, Branch of Distribution, 1200 South Eads Street, Arlington, Virginia 22202 will also provide a free state topographic index map. The procedure for purchasing topographic maps directly from the U.S. Geological Survey is explained on the index map. Topographic maps can also be obtained from most state geological surveys and private map services.

Public museums with geologic displays and geology departments associated with a nearby college or university are other sources of information on local geology. Libraries associated with these institutions are the best places to find geological publications.

The state surveys and some geology departments will periodically host public field trips. In addition, there are several organizations in this region that conduct field trips. These include the Tri-State Field Conference, the Lake Superior Institute, and the Michigan Basin Geological Society. Information on the programs of these organizations is available from your nearest geology department or state geological survey. There are also numerous active associations of highly talented amateur geologists ("rock hounds") throughout this region. These organizations are usually involved in a busy schedule of field trips and educational programs. For information on the organization nearest you write: Midwest Federation of Mineralogical Societies, P.O. Box 1130, Des Moines, Iowa 50311.

The following are specific references we cited:

Alden, William C., 1918, The Quaternary geology of southeastern Wisconsin: U.S. Geological Survey Professional Paper 106, 350 p.

Alden, William C., 1932, Glacial geology of the central states: International Geological Congress Guidebook 26, Excursion C-3, 54 p.

Black, Robert F., 1964, The physical geography of Wisconsin; *in* 1964 Wisconsin Blue Book: State of Wisconsin, p. 171-177.

Dalziel, I.W.D., and R.H. Dott, Jr., 1970, Geology of the Baraboo district, Wisconsin: Wisconsin Geological and Natural History Survey Information Circular No. 14, 163 p.

Dorr, John A., and Donald F. Eschmann, 1971, Geology of Michigan: The University of Michigan Press, Ann Abor, 476 p.

Fenneman, N.M., 1938, Physiography of eastern United States: McGraw-Hill, New York, 714 p.

Frye, John C., H.B. Willman and Robert F. Black, 1965, Outline of glacial geology of Illinois and Wisconsin *in* The Quaternary of the United States, edited by H.E. Wright Jr. and David G. Frey: Princeton University Press, p. 43-62.

Gorder, Lyle, 1975, The coastal zone in northeastern Wisconsin; *in* Landscapes of Wisconsin, edited by Barbara Zakrzewska-Borowiecki: Association of American Geographers, Washington D.C., p. 52-68.

Harbaugh, John W., 1968, Stratigraphy and geologic time: Wm. C. Brown Company Publishers, Dubuque, 113 p.

Holmes, Arthur, 1965, Principles of physical geology: Ronald Press, New York, 1288 p.

Kulp. J.L., 1961, Geologic time scale: Science, vol. 133, p. 1105-1114.

Martin, Lawrence, 1932, The physical geography of Wisconsin: Wisconsin Geological and Natural History Survey Bulletin 36, 608 p. Reprinted 1965, University of Wisconsin Press.

Ostrom, M.E., 1967, Paleozoic stratigraphic nomenclature for Wisconsin: Wisconsin Geological and Natural History Survey, Information Circular 8.

Schwartz, George M., and George A. Thiel, 1973, Minnesota's rocks and waters: The University of Minnesota Press, Minneapolis, 366 p.

Sims, P.K., and G.B. Morey, editors, 1972, Geology of Minnesota: Minnesota Geological Survey, 632 p.

U.S. Geological Survey, 1970, The national atlas of the United States of America: U.S. Dept. of Interior, Geological Survey, Washington, D.C., 417 p.

Van Schmus, W.R., E.M. Thurman and Z.E. Peterman, 1975, Geology and Rb-Sr chronology of Middle Precambrian rocks in eastern and central Wisconsin: Bulletin of Geological Society of America, vol. 86, p. 1255-1265.

Willman, H.B., 1971, Summary of the geology of the Chicago area: Illinois Geological Survey Circular 460, 77 p.

Wolosin, Carl A., 1972, Seismic investigations and petroleum evaluation of central Lake Michigan: Unpublished Master's thesis, University of Wisconsin-Milwaukee, 149 p.

GLOSSARY

For your convenience, more than 300 geologic terms used in this book are defined below. In many instances the definition is followed by a specific reference to the text where the term is illustrated by a figure, or defined in more detail.

Acidic igneous rock: An intrusive or extrusive igneous rock composed predominantly of minerals with a high silica content. Such a rock is usually light-colored, and has an average specific gravity.

Acidic water: Water that contains sufficient hydrogen ions to have a pH of less than 7. Such water will dissolve limestone and dolomite.

Algal dolomite: A dolomite that preserves fossil evidence of algae. See p. 30.

Alluvial fan: A gently sloping mass of loose clay, sand, and rock in the general shape of a fan. Such features form where a rapidly flowing stream or river issues from a highland onto a plain or broad valley.

Altonian: A glacial substage that involved an ice advance during the initial half of the Wisconsinan Stage. See p. 62 and Table VI on p. 63.

Amphibole: A group of iron, magnesium, aluminum silicate minerals that also contain calcium and sodium. These minerals are generally dark-colored and elongate. They are common in intermediate and basic igneous rocks, and in many metamorphic rock types. See p. 6.

Amygdaloidal: Refers to extrusive igneous rocks containing numerous gas cavities (vesicles) that were later infilled with secondary minerals. See amygdules and vesicles below.

Amygdule: A gas bubble in an extrusive igneous rock that is infilled with minerals. See p. 3.

Andesite: A dark-colored, fine-grained, generally extrusive rock composed of plagioclase with intermediate composition, quartz, and either biotite, hornblende, or pyroxene. See p. 6.

Angular unconformity: A type of unconformity in which the underlying (older) rocks meet the overlying (younger) rocks at an angle. See p. 14.

Animikean: A series of predominantly metamorphosed sedimentary rocks that are middle Precambrian in age. See Fig. 2-7 on p. 30.

Anorthosite: An intrusive, plutonic, igneous rock composed almost entirely of calcium-rich plagioclase feldspar. See p. 5.

Anticline: An arching, or upfolding in which the oldest rocks are in the center of the structure. See p. 22, Fig. 1-14.

Aquifer: A rock or sediment layer that is saturated with water, and contains enough void space to yield significant quantities of ground water to wells and/or springs.

Arch: An anticlinal upwarp that is usually of regional scale. See anticline above, Fig. 1-3 on p. 4, and p. 5.

Argillaceous: Describes the presence of significant amounts of clay-size particles or clay minerals.

Arkose: A sandstone containing abundant feldspar grains. See Table II on p. 8.

Arthropod: A group of invertebrates characterized by jointed appendages and segmented bodies. See p. 180 and Fig. 7-5 on p. 181.

Ash fall: An accumulation of fine volcanic particles that settled to the earth after an explosive eruption.

Atoll: A ring-shaped group of coral reefs that developed into low coralline islands or islets, which enclose a lagoon.

Augite: An iron, magnesium, aluminum silicate mineral that also contains calcium and sodium. This mineral is usually found as black, equant-shaped crystals in basic igneous rocks, and in some metamorphic rocks.

Banded: Describes a metamorphic rock containing thin, alternating layers of light and dark minerals. See Table III on p. 9.

Basal conglomerate: A conglomerate that develops on an erosional surface. As such, it is the initial deposit overlying an unconformity. See p. 14.

Basalt: An extrusive, basic igneous rock. This rock type is rich in iron and magnesium, and deficient in silica. Hence, it is fine-grained, dark-colored, and relatively heavy. See p. 1.

Basic igneous rock: An intrusive or extrusive igneous rock that is rich in iron, magnesium, and/or calcium, and deficient in silica. Such rocks are usually dark-colored, and have a relatively high specific gravity. See p. 5.

Basin: A synclinal downwarp, either circular or elliptical in outline, in which the rock units dip inward in all directions. Such features are often of regional scale. See syncline below, Fig. 1-3 on p. 4, and p. 5.

Batholith: A large, intrusive, igneous body, generally composed of medium to coarse-grained acidic rock. See p. 1 and Fig. 1-2 on p. 3.

Bathymetric: Refers to the measurement of water depths, and the mapping of the bottom of water bodies such as lakes, seas, and oceans.

Baymouth bar: A spit that grows completely across the entrance to a bay. See spit below, p. 13, and Fig. 6-8 on p. 157.

Bedding plane: The generally planar surface that separates layers of stratified rock.

Bedrock: Rock that is exposed at the surface, or underlies soil or other unconsolidated material in a given area.

Biotite: A dark-green, brown, or black, platy, mica mineral that is common in many igneous and metamorphic rocks. See mica below and p. 6.

Block diagram: A three-dimensional diagram that combines a surface view with two vertical cross sections at right angles to each other. See p. 21 and Fig. 1-14 on p. 22.

Boulder train: A series of glacial erratics that is distributed outward for some distance from the same bedrock source. The dispersal pattern of such erratics helps define the direction of glacial ice movement. See erratic below and p. 56.

Brachiopod: A marine, bivalved animal somewhat similar to a clam, but with opposing valves that are distinctly different. See Fig. 7-4 on p. 178 and p. 179.

Breccia; Brecciated: A rock composed of broken and angular rock fragments that are cemented together with material of similar or different composition. Brecciated is the term describing the characteristic texture of such a rock.

Bryozoan: A colonial, marine invertebrate somewhat like a coral. However, a bryozoan has a U-shaped alimentary canal, while a coral does not. See Fig. 7-4 on p. 178 and p. 179.

Butte: An isolated, flat-topped hill with relatively steep slopes. Such features are usually capped with a resistant layer of rock, and represent an erosional remnant of a surface which once had greater areal extent.

Calcareous: An adjective used to describe a rock that contains a considerable amount (up to 50%) of calcium carbonate (calcite).

Calcium plagioclase: A plagioclase feldspar that contains more calcium than it does sodium. Such feldspars are common in basic igneous rocks, and in some metamorphic rock types. See plagioclase feldspar below.

Calyx: The cup-shaped, oval assemblage of calcite plates that forms the head of a crinoid or cystoid. See p. 180 and Fig. 7-5 on p. 181.

Cambrian: The earliest geologic period in the Paleozoic Era. This period marked the beginning of an extensive fossil record. It is sometimes called the age of trilobites. See Table IV on p. 17.

Cap rock: A resistant rock layer that protects less resistant rocks below it from erosion. See Fig. 4-1 on p. 97.

Carboniferous: The combined Mississippian and Pennsylvanian geologic periods. See Table IV on p. 17.

Castellated: An adjective used to describe a high-standing topographic feature that includes towering, or battlement-like, rock exposures.

Cave: A tunnellike opening in bedrock that is usually formed by the solution of limestone or dolomite by migrating acidic waters. See Fig. 5-2 on p. 113.

Cement: Minerals that form in void spaces within rock or sediment. Cementation is a rock-forming process that serves to bind mineral grains together into a rigid, coherent mass. Common cements include calcite, silica, dolomite, and iron oxides.

Cenozoic: An era of geologic time that spans the interval from the end of the Mesozoic to the present. See Table IV on p. 17.

Central Lowlands: The northern part of the stable interior of the North American continent. As such, it includes all of our region south of the Northern Highland. See p. 23 and Fig. 2-2 on p. 25.

Central Plain: A lowland area developed on Cambrian sandstone around the southern margin of the Northern Highland. See p. 72, Fig. 3-1 on p. 73, and Chapter 4.

Cephalon: The head portion of a trilobite, which includes several fused segments and the eyes and mouth. See p. 180.

Cephalopod: A marine mollusk that has a well-defined head with tentacles. Extinct forms had hollow, segmented shells which ranged in shape from straight to coiled. See p. 179, 180, and Fig. 7-5 on p. 181.

Chatter marks: A series of small, chiplike scars on a bedrock surface that were made by rocks contained in the base of a glacier.

Chert: A hard, dense, mass of silica that consists of individual quartz crystals that are too small to be seen, even with the aid of a microscope (cryptocrystalline).

Chitin: A resistant, organic compound that forms the exoskeletons of insects and some invertebrates.

Chlorite: A group of iron, magnesium, aluminum silicates that are platy and usually greenish in color. Chlorite is common in low-grade metamorphic rocks. See p. 9.

Clastic: An adjective used to describe a rock or sediment primarily composed of broken rock fragments and/or mineral grains derived from preexisting rocks.

Clay minerals: Finely-crystalline, hydrous silicates of aluminum, iron, or magnesium, that may also contain potassium, sodium, etc. They are usually formed by the alteration or weathering of rock-forming silicate minerals such as feldspars, pyroxenes, and amphiboles.

Cleavage: The tendency of a rock to break repeatedly along a smooth surface at some angle to bedding. Cleavage also refers to the ability of many minerals to part along preferred planes that reflect the internal crystalline structure. See p. 9 and p. 120.

Coelenterate: A group of marine animals that includes the corals, jellyfish, and sea fans. See p. 177, Fig. 7-4 on p. 178, and p. 179.

Conformable: A term used to describe an undisturbed sequence of bedded rock units. Such a relationship implies that deposition continued without interruption, and adjacent rock units often grade into each other.

Conglomerate: A coarse-grained, clastic, sedimentary rock composed of at least 25% rounded to subangular particles larger than 2 mm, in a finer matrix.

Conodont: A microscopic, phosphatic structure from an animal of unknown affinity that is useful in age-dating and correlation. See p. 182.

Contact: The planar or irregular surface that separates two rock units.

Continental interior: The stable, central portion of the North American continent. See Fig. 2-1 on p. 24.

Continental plates: Those parts of the crust of the earth that are the relatively light (specific gravity about 2.7), complex, rock masses that form the individual continents. Each of these continental plates can move independently of the others, and all of them overlie a denser layer of the crust which forms the bedrock of the deep ocean basins.

Coral: An attached, bottom-dwelling, marine, invertebrate animal. Colonial corals were important reef-builders during part of the Paleozoic. See p. 177, Fig. 7-4 on p. 178, and p. 179.

Correlation: The determination of the spatial position and/or the time of occurrence of one geologic feature, or event, in relation to others. See Fig. 1-13 on p. 20.

Crest: The line defined by connecting the maximum elevations of a given bed in an anticlinal structure. Hence, it defines the highest part of the structure. Also, a line connecting the highest points of a landform, from which the surface slopes downward in opposite directions.

Cretaceous: The last geologic period in the Mesozoic Era. See Table IV on p. 17.

Crevasse fill: A short, linear ridge of water-washed sand and gravel that was deposited in a crack (crevasse) on a stagnating glacier. See p. 58.

Crinoid: A marine echinoderm with a budlike calyx, generally composed of calcareous plates. Most fossil forms were attached to the sea floor by a calcareous stem, and had gathering arms to secure food. See p. 180 and Fig. 7-5 on p. 181.

Cross-bedding: An arrangement of layering within a stratified deposit where the internal bedding is inclined at an angle to the primary bedding planes. Such a feature forms in sediments deposited by wind or water currents, and the direction of these currents can be determined by examining the orientation of the cross-bedding.

Crosscutting relationships, Law of: The principle which states that a rock unit is older than any rock body which penetrates into, or cuts across it. See p. 14 and Fig. 1-10 on p. 15.

Crustacean: An arthropod characterized by the presence of two pairs of antennae on its head. See arthropod above, p. 180, and Fig. 7-5 on p. 181.

Crustal rebound: The gradual rise of a landmass in response to the retreat of continental ice sheets.

Cuesta: A persistent ridge with a gentle slope on one side and a steep slope on the other. Such features reflect variations in the erosional resistance of gently-dipping, adjacent rock units. See Fig. 4-1 on p. 97.

Cystoid: An extinct, stemmed echinoderm with a head (calyx) composed of numerous, irregularly arranged, calcareous plates. See p. 180 and Fig. 7-5 on p. 181.

Delta: A low-lying, fan-shaped deposit formed where a river drops its sediment load upon entering a body of water, such as a lake or ocean.

Dendritic: Describes a stream drainage pattern that resembles the branching habit of trees. This is the pattern that develops when the underlying bedrock is homogeneous, or deeply mantled by unconsolidated material.

Deposition: The precipitation, settling, dropping, or laying-down of minerals and/or organic matter to form a sedimentary accumulation.

Devonian: The fourth geologic period of the Paleozoic Era. Fossil plants and fish made a dramatic appearance during this period. See Table IV on p. 17.

Dike: A tabular igneous body that cuts across earlier formed rocks. See p. 1 and Fig. 1 on p. 3.

Diorite: A plutonic igneous rock characteristically composed of dark-colored amphibole and intermediate plagioclase. In some instances, minor amounts of pyroxene or quartz are present. See p. 6 and Table I on p. 7.

Dip: Angle of inclination of a rock surface from the horizontal. By definition, dip is at right angles to strike. See strike below and p. 21.

Dolomite: A calcium, magnesium carbonate mineral similar to calcite in appearance. Also, a sedimentary rock containing more than 50% of the mineral dolomite. See p. 3.

Dolomitization: The process whereby limestone (calcite) is completely, or partly, converted to dolomite, or dolomitic limestone, when magnesium partially replaces calcium to form the mineral dolomite (calcium magnesium carbonate).

Dome: An anticlinal uplift, either circular or elliptical in outline, in which the rock units dip away in all directions.

Drift: See glacial drift below.

Driftless Area: A region that includes southwestern Wisconsin, and immediately adjacent parts of Illinois, Iowa, and Minnesota. It was surrounded, but apparently not covered, by continental ice sheets during the Pleistocene Epoch. See Fig. 2-28 on p. 55.

Drumlin: A streamlined, oval, elongate mound molded from till by glacial ice. Such features have a steep slope that faced the direction of ice advance, and a long axis parallel to the direction of ice flow. See p. 54 and Fig. 2-30 on p. 57.

Dune: A mound or ridge of wind-deposited sand or silt. See p. 154, Fig. 6-7 on p. 155, and p. 156.

Eastern Ridges and Lowlands: The province that forms the eastern part of our area. It is formed on Paleozoic rocks and is characterized by a series of relatively resistant ridges alternating with lowland areas. The entire province is heavily veneered by glacial deposits. See p. 72, Fig. 3-1 on p. 73, and Chapter 6.

Echinoderm: A group of marine animals possessing five-fold, radial symmetry, and generally an exterior skeleton of interlocking, calcium carbonate plates. Echinoderms include cystoids, blastoids, starfish, crinoids, sea urchins, and sea cucumbers. See p. 180 and Fig. 7-5 on p. 181.

End moraine: Unsorted, glacial debris (till) deposited in a definable ridge (or ridges) at the terminus of a glacier. Such accumulations may mark the maximum extent of a glacier (terminal moraine), or they may be deposits formed during a spasmodic retreat of an ice sheet (recessional moraines).

Eocene: An epoch in the lower Tertiary. See Table IV on p. 17.

Epidote: A calcium, iron, aluminum silicate mineral that is generally greenish in color, and massive. It is usually a product of low-grade metamorphism.

Epoch: A time-defined subdivision of a geologic period. See Table IV on p. 17.

Era: The largest division of the geologic time scale. See Table IV on p. 17.

Erosion: The process by which materials on the earth's surface are loosened, broken-up, dissolved, and transported from one place to another.

Erratic: A fragment of bedrock picked up by glacial ice, and carried some distance before deposition. See p. 54, p. 55, and Fig. 2-31 on p. 57.

Escarpment: A long, continuous cliff or relatively steep slope facing in one general direction. It commonly marks the position of a resistant rock unit. See Fig. 6-1 on p. 140 for an illustration of the Silurian escarpment.

Esker: A sinuous, narrow ridge of sand and gravel formed by a stream that flowed on, within, or under glacial ice. See p. 58 and Fig. 2-34 on p. 59.

Exposure: Bedrock that is visible at the surface. Such occurrences may be natural or manmade.

Extrusive: Igneous material that is ejected from the interior of the earth and accumulates on the surface.

Farmdalian: The early Wisconsinan interglacial interval between the Altonian and Woodfordian glacial advances. See p. 62 and Table VI on p. 63.

Fault: A surface or zone of rock fracture where there has been sufficient movement to produce measurable displacement. See Fig. 1-14 on p. 22.

Fault plane: The movement surface of a fault. See fault above and Fig. 1-14 on p. 22.

Faunal succession: The observable sequence of change in animal life through geologic time. See p. 15.

Feldspar: A group of aluminum silicate minerals that generally contain potassium, sodium and/or calcium. Those containing potassium are orthoclase or microcline. The ones rich in sodium and/or calcium are the plagioclase feldspars. See orthoclase, microcline, and plagioclase feldspar below.

Ferromagnesian: Refers to minerals or rocks that are rich in iron and magnesium.

Fissility: Describes the ability of a rock to split or part along closely spaced planes that are essentially parallel to bedding. Shale is a good example of a rock with this property.

Fissure: A fracture or crack in rock with pronounced separation.

Flood plain: The expanse of relatively flat land adjacent to a river that is subjected to periodic flooding. See Fig. 1-6 on p. 11.

Floral succession: The observable sequence of change in plant life through geologic time. See p. 15.

Flow banding: Alternating layers with differing mineralogy and/or texture that form when igneous rocks crystallize from moving magma or lava. In most instances, the minerals are aligned parallel to the direction of flow.

Flux: As we have used it, a substance that is added to molten metals in a smelter to remove nonmetallic impurities.

Fold: A deformational feature produced by the bending of rocks. See p. 21 and Fig. 1-14 on p. 22.

Foliation; Foliated: The rock property produced by the preferred alignment of minerals, such as the micas (muscovite, biotite), within a metamorphic rock. This causes the rock to part along planes of weakness which parallel the pattern of mineral orientation. A foliated rock is one with a well-developed foliation. See p. 9.

Formation: The basic unit in the description or organization of rock sequences in a given area. A formation is a body of rock that is thick enough to be mapped, and definable by lithologic features that distinguish it from overlying and underlying rock units. See p. 19.

Fossil: Any evidence of past life with geologic antiquity. See p. 176.

Fracture: A general term for any break in a rock. As such, fractures would include cracks, fissures, joints, and faults. However, the term usually implies an irregular break without displacement of the adjacent walls.

Gabbro: A heavy, dark-colored, coarse to medium-grained, basic, plutonic, igneous rock primarily composed of calcium-rich plagioclase and augite. See p. 1 and Table I on p. 7.

Galena: A very heavy, grey to black, lead sulphide mineral that forms cubic crystals. It is a common lead ore in the Upper Mississippi Valley Lead-Zinc District. See Fig. 7-2 on p. 175.

Garnet: A group of calcium, magnesium, iron, aluminum silicate minerals that often form red, but occasionally green or brown, crystals. Garnets are common in some metamorphic rocks, but they also occur occasionally in igneous rocks.

Gastropod: A group of mollusks that includes the marine and nonmarine snails. See Fig. 7-4 on p. 178 and p. 179.

Geologic time scale: An arbitrary, chronologic arrangement of geologic events that provides a relative time sequence. More recently, absolute age determinations based on radiometric dating have been added to the time scale, so that the duration and/or age of subdivisions in years are now known. See Table IV on p. 17.

Glacial drift: A general term for all types of glacial deposits. See p. 52.

Glacial lake: A large water body formed by glacial meltwater impounded by an ice front and/or by drift ridges. See p. 59.

Glacial milk: Turbid, whitish, glacial meltwater containing finely-ground silt and clay in suspension. See p. 57.

Glacial outwash: Stratified sand and gravel deposited by meltwater streams at or beyond the margin of a glacier. See p. 57 and Fig. 2-32 on p. 58.

Glacier: A large mass of ice formed by the progressive accumulation, compaction, and recrystallization of snow. Such a mass moves downslope, or outward from the center of accumulation due to gravity flow. See Fig. 1-7 on p. 12.

Glauconite: A greenish, sand-size, potassium, iron, magnesium, aluminum silicate mineral that is common in many Cambrian sandstone units in our region. See p. 40.

Gneiss: A dense, banded, high-grade metamorphic rock with a composition that is similar to granitic igneous rocks. See p. 4 and Table III on p. 9.

Grade: A term that is used to describe a river in which erosion and deposition are in balance. The term is also used to describe the quality of ore (i.e., high-grade, low-grade).

Granite: A medium to coarse-grained, plutonic, acidic igneous rock containing silica, potassium, and sodium as its major constituents. The common minerals present include quartz, potassium feldspar (orthoclase and/or microcline), sodium-rich plagioclase, muscovite, biotite, and hornblende. See p. 1 and Table I on p. 7.

Graptolite: An extinct, colonial, marine animal represented in the fossii record by tiny, chitinous, cuplike structures that are grouped along a narrow, straight, or branched rod. See p. 180 and Fig. 7-5 on p. 181.

Greenstone: A massive, dark-greenish, chlorite-bearing rock that forms from the metamorphism of basic igneous rocks. See p. 8.

Greywacke: A sandstone largely composed of sand-size fragments of dark rock. Such rocks usually contain appreciable amounts of silt and clay-size particles. See Table II on p. 8.

Groin: A barrier that projects into a lake or ocean at right-angles to the shoreline, to encourage on-shore deposition of sand by deflecting longshore currents into deeper water. See p. 158 and Fig. 6-10 on p. 159.

Ground moraine: Glacial till deposited as an uneven veneer to form an irregular, rolling surface. See p. 54.

Ground water table: The nonhorizontal, fluctuating, upper surface of the zone where all the spaces within rock or sediment are water-filled. See Fig. 1-5 on p. 10.

Gypsum: A soft, calcium sulfate mineral that forms during the evaporation of sea water. See p. 7.

Halite: A mineral, composed of sodium chloride, that is better known as rock salt. Halite is the common mineral that forms during the evaporation of sea water. See p. 7.

Hematite: An iron oxide mineral that is an important ore of iron. See p. 89.

Hornblende: A black, elongate, prismatic, calcium, sodium, magnesium, iron, aluminum silicate mineral that is an important constituent of acidic and intermediate igneous rocks, and some metamorphic rocks.

Hydrologic cycle: The cyclical concept that interrelates the atmospheric, surface, and subsurface water systems of the world. See Fig. 1-5 on p. 10.

Igneous rocks: Rocks that crystallize from a melt (magma) within, or at the surface of the earth. See p. 1 and Table I on p. 7.

Inclusions, Law of: The concept that enclosing rock is younger than any included material. See p. 14.

Inlier: An isolated exposure of bedrock surrounded by rocks of younger age. See p. 77.

Interlobate moraine: A moraine that formed between two adjacent lobes of glacial ice. See p. 54 and Fig. 2-28 on p. 55.

Intermediate igneous rocks: Igneous rocks with a silica content that is intermediate between basic and acidic igneous rocks. Such rocks may be intrusive or extrusive. They contain plagioclase feldspars with appreciable amounts of both calcium and sodium, some quartz, and various other minerals. See p. 6 and Table I on p. 7.

Inter-reef: Dense, thin to medium-bedded, sparsely-fossiliferous rocks that formed between reefs. See p. 170 and Fig. 6-18 on p. 171.

Intrusive igneous rocks: Igneous rocks that cross-cut (intrude) preexisting rocks. See Fig. 1-2 on p. 3.

Iron formation: A siliceous, typically laminated or thin-bedded, sedimentary rock that contains more than 15% iron. See p. 89.

Jaspilite: An iron formation with alternating silica-rich and iron-rich layers; generally reddish in color. See p. 89.

Jetty: A man-made structure that extends outward from the shore into a standing body of water to control waves or currents.

Joint: A relatively smooth surface of fracture or parting in a rock without displacement. In a given area, two favored directions of jointing are commonly recognized. The orientation of these is referred to as the joint pattern.

Jurassic: The middle geologic period in the Mesozoic Era. See Table IV on p. 17.

Kame: In our area, most kames are conical hills of sand and gravel deposited beneath the ice by glacial meltwater that swirled downward through cylindrical holes or shafts. See p. 58; including Fig. 2-33.

Keewatin: The oldest rock series exposed in our region. As such, they are metamorphosed lava flows and gneissic rocks of Early Precambrian age. See Fig. 2-7 on p. 30.

Kettle: A depression formed by the melting of a large, detached block of glacial ice that was partially, or completely buried in glacial drift. See p. 60.

Kettle lake: A lake within a kettle that is deep enough to intersect the water table; usually found in pitted outwash deposits. See p. 60; including Fig. 2-35.

Keweenawan: A series of lava flows and sedimentary rocks of Late Precambrian age. See Fig. 2-11 on p. 33.

Knob and swale topography: An undulating glacial landscape with a disordered assemblage of kames (knobs) and irregular depressions, some of which are kettles. See kames and kettles above.

Lamination: Extremely fine, sedimentary layering that is common in shale and very fine-grained sandstone.

Lava; Lava flow: An outpouring of magma onto the earth's surface from a vent or fissure. It crystallizes into a body of igneous rock that extends laterally for some distance. See Fig. 1-2 on p. 3.

Limb: The flank, or side of an anticlinal or synclinal fold.

Limestone: A sedimentary rock, formed from organic or inorganic processes, that is composed of more than 50% calcite (calcium carbonate). See p. 3 and Table II on p. 8.

Limonite: A yellowish-brown, earthy, iron oxide mineral that was once mined as an ore of iron in our area. See p. 89 and p. 193.

Lithologic: The overall characteristics of a rock; including the mineralogic composition, texture, color, and structure.

Lobe: A rounded, tonguelike projection of glacial ice that extends beyond the main mass of a continental ice sheet.

Loess: Unconsolidated, windblown silt. In our area, loess refers to fine-grained, glacial outwash sediments that were picked up by the wind and deposited over much of the Upper Mississippi Valley. See p. 59.

Longshore current: A wave-generated current that flows in shallow water parallel to the shoreline of a lake or ocean. See Fig. 1-9 on p. 13.

Magma: Molten material, composed of ions, formed by the localized fusion of rock in the crust. See p. 1.

Marble: Metamorphosed limestone. See p. 8.

Marl: A poorly consolidated, earthy substance consisting of a mixture of clay minerals and calcite (calcium carbonate). In our area, marls are fresh water deposits of Pleistocene or Recent age.

Massive: A term used to describe a rock that is homogeneous and essentially without bedding or other internal features, such as laminations, banding, foliation, etc.

Mastodon: An extinct, large, tusked mammal that was a relative of the elephant, and roamed the earth during the Pleistocene. See p. 182.

Meander: One of a series of somewhat regular, sinuous curves along the course of a stream or river. See p. 11.

Melt: A liquefied mass (magma) of fused rock. See magma above and p. 1.

Merostome: A small, segmented animal somewhat similar in appearance to a trilobite, but biologically more complex. See p. 180.

Mesa: An isolated, level, relatively large landform entirely surrounded by steep slopes. It is similar to a butte (see above), but more extensive in surface area.

Mesozoic: The third era in the geologic time scale. See Table IV on p. 17.

Metamorphosed; Metamorphic rock: Any change in the composition, texture, internal structure, etc., of a rock produced by temperature, pressure, or migrating fluids. The rock that results from such changes. See p. 3, p. 4, and Table III on p. 9.

Mica; Micaceous: A group of soft minerals that are generally hydrous, potassium, sodium, calcium, magnesium, iron, aluminum silicates which readily split into thin sheets or flakes. Some micas, such as muscovite and biotite, are common rock-forming minerals in granitic igneous rocks and some metamorphic rocks. Micaceous is used to describe rocks that contain abundant mica, or for minerals, such as some chlorites, that split into thin flakes like the micas.

Micritic: A descriptive term for very fine-grained, crystalline calcite that may be chemically precipitated calcium carbonate mud or finely-ground fossil debris. Such material is common as matrix in carbonate rock, and in some instances it may form the entire rock. See Table II on p. 8.

Microcline: A clear to milky, white, red, or green, potassium, aluminum silicate that forms rectangular-appearing crystals in acidic igneous rocks, and some metamorphic rocks. This is one of the two common potassium feldspars; the other is orthoclase.

Mineral: A solid, natural-occurring, inorganic element or compound with a definite chemical composition and an internal crystalline structure. See p. 173.

Mineralogy: The study of the occurrence, composition, and properties of minerals.

Miocene: The fourth epoch in the Tertiary geologic period. See Table IV on p. 17.

Mississippian: The fifth geologic period of the Paleozoic Era. See Table IV on p. 17.

Mollusk: A phylum of invertebrate animals that includes the gastropods (snails), cephalopods, and pelecypods (clams). See Fig. 7-4 on p. 178, p. 179, and Fig. 7-5 on p. 181.

Monadnock: An isolated, resistant, bedrock ridge or hill that rises above the level of an extensive erosional surface. See p. 77.

Moraine: The general term for an accumulation of unsorted, unstratified, glacial drift. See p. 54.

Moulin kame: A conical hill of water-washed sand and gravel that was deposited by meltwater flowing down through a hole in glacial ice. See Fig. 2-33 on p. 58.

Mound: In our area, an isolated, rounded, bedrock hill which is an erosional remnant (outlier), that rises above the general level of the surrounding countryside. See Fig. 4-1 on p. 97 and p. 127.

Mudstone: A sedimentary rock that is massive in character, and composed of a mixture of sand, silt, and clay (mud). As such, it is different from shale, which readily parts along closely spaced planes parallel to bedding.

Muscovite: A generally clear mica consisting of hydrous, potassium, aluminum silicate. Muscovite is common in acidic igneous rocks, metamorphic rocks, and as a clastic grain in some sedimentary rocks. See p. 6 and mica above.

Native copper: Metallic copper; once an important source of this metal in the Keweenaw Peninsula of northern Michigan. See Fig. 7-1 on p. 174.

Natural bridge: An isolated arch of bedrock created by erosional processes. See Fig. 5-17 on p. 131.

Nodular: Characterized by rounded or tuberous lumps that may occur alone or as aggregations.

Normal fault: A break in a rock sequence where the displacement is such that movement was down for rocks above the inclined fault plane relative to those below it. See Fig. 1-14 on p. 22.

Northern Highland: The province that forms the northern part of our region. It is a highland area of predominantly Precambrian bedrock, that is veneered by Pleistocene glacial deposits. As such, it is a southward extension of the Canadian shield. See Fig. 3-1 on p. 73 and Chapter 3.

Obsidian: A glassy, dark rock formed by the very rapid cooling of magma. See p. 6 and Table I on p. 7.

Oligocene: The third epoch in the Tertiary geologic period. See Table IV on p. 17.

Olivine: An olive-green, glassy-appearing, iron,

magnesium silicate that forms spherical-appearing crystals in basic igneous rocks.

Oolite: A small, pelletlike grain composed of thin, concentric layers of calcium carbonate or iron compounds which formed around a nucleus, such as a mineral or fossil fragment. See p. 47.

Ordovician: The second geologic period in the Paleozoic Era. See Table IV on p. 17.

Ore: A mixture of minerals and rock that can be mined at a profit.

Original horizontality, Law of: The principle which states that most sedimentary rocks were deposited as flat-lying units parallel to the earth's surface. See p. 21.

Orogenic cycle: The concept that envisions a cyclic sequence of events characterized by uplift, erosion, deposition, subsidence, metamorphism, and possibly melting of a rock succession in a given area. See Fig. 1-3 on p. 4 and p. 5.

Orogeny: An intense interval of structural deformation and uplift that often is accompanied by large-scale igneous intrusions. Such events form major mountain systems. See p. 5.

Orthoclase: A generally colorless to gray, potassium, aluminum silicate mineral that forms rectangular-appearing crystals in acidic igneous rocks, and some metamorphic rocks. This is one of the two common potassium feldspars; the other is microcline.

Ostracod: A small to microscopic, bivalved crustacean resembling a navy bean. See p. 180 and Fig. 7-5 on p. 181.

Outcrop: A natural or man-made bedrock exposure.

Outlier: An isolated bedrock exposure that is entirely surrounded by older rocks. Mounds, mesas, and buttes are examples of outliers. See p. 77 and Fig. 4-1 on p. 97.

Outwash plain: A relatively flat expanse of sand and gravel that was deposited by glacial meltwaters beyond the ice margin. See p. 59.

Paleocene: The oldest epoch in the Tertiary geologic period. See Table IV on p. 17.

Paleozoic: The second era of the geologic time scale. See Table IV on p. 17.

Peat: An unconsolidated accumulation of plant remains deposited in a water-saturated environment such as a bog. When dried, it can be burned as a low-grade fuel. See p. 190.

Peccary: A piglike animal that inhabited this area during the Pleistocene. See p. 182.

Pegmatite: A very coarsely-crystalline, intrusive, igneous, dike rock that contains abundant quartz, potassium feldspar, and muscovite. Pegmatites represent a very late stage in the cooling history of a magma, and they may contain relatively rare minerals.

Pelecypod: Bivalved, aquatic mollusks, which are commonly called clams. See Fig. 7-4 on p. 178 and p. 179.

Pennsylvanian: The sixth geologic period of the Paleozoic Era. See Table IV on p. 17.

Peridotite: A coarse-grained, plutonic, basic igneous rock that is rich in augite and olivine. See p. 5.

Period: The subdivision of a geologic era. See Table IV on p. 17.

Permafrost: Permanently frozen ground. See p. 51.

Permeable: The capacity of a sediment or rock to allow fluids (water) to move readily through interconnected void spaces.

Permian: The seventh, and last geologic period in the Paleozoic Era. See Table IV on p. 17.

Phosphatic: Describes a rock that contains phosphate minerals.

Phyllite: An argillaceous rock that is intermediate in metamorphic grade between slate and schist. See argillaceous above and footnote on p. 120.

Phylum: The primary subdivision of the plant and animal kingdoms.

Pillow lava: A rock structure in basalt which resembles a bed pillow. Such features form when lava is extruded under water. See Fig. 2-5 on p. 28.

Plagioclase feldspar: A series of common, rock-forming, aluminum silicate minerals that contain variable amounts of sodium and calcium. See p. 5.

Plateau: A relatively flat, extensive highland region.

Pleistocene: The first epoch of the Quaternary geologic period. This interval is often referred to as the Ice Age. See Table IV on p. 17 and Table VI on p. 63.

Pliocene: The fifth epoch of the Tertiary geologic period. See Table IV on p. 17.

Plug: A near-vertical, pipelike body of igneous rock that probably formed the conduit to a volcano.

Plutonic rocks: Igneous rocks that crystallized below the surface of the earth. See p. 1 and Fig. 1-2 on p. 3.

Porous: Describes the presence of numerous void

spaces within a sediment or rock. These may include: spaces between individual mineral grains, open fractures, gas bubbles within lava flows, solution cavities in limestone, etc.

Porphyry: An igneous rock that contains large crystals surrounded by a finer grained matrix. See Table I on p. 7.

Potassium feldspar: Potassium, aluminum silicate minerals that are common rock-formers. These include both orthoclase and microcline. See orthoclase and microcline above.

Pothole: A cylindrical depression eroded into bedrock by pebbles and cobbles that were swirled around on the bottom of a swiftly flowing river by eddy currents. See Fig. 3-9 on p. 87.

Precambrian: The oldest, major division in the geologic time scale; equivalent to about 90% of geologic time. See Table IV on p. 17.

Precipitate: A deposit that forms from a chemical reaction, and from the evaporation of sea water (evaporites) or ground water.

Province: An area with similar physical characteristics, such as bedrock geology and/ or topography. See p. 71.

Pygidium: The tail portion of a trilobite. See p. 180.

Pyroxene: A family of dark, calcium, sodium, magnesium, iron, aluminum silicate minerals that form equant-appearing crystals in basic igneous rocks and high-grade metamorphic rocks.

Quartz: A generally glassy or milky-appearing mineral composed of crystalline silica (silicon dioxide). This is one of our commonest minerals, and it is an important constituent in diverse rock types that include: sandstone, conglomerate, shale, acidic to intermediate igneous rocks, quartzite, schist, and gneiss. It also occurs as veins, and forms prismatic crystals within void spaces in many rocks.

Quartzite: A metamorphic rock composed of sand-size quartz grains which are either so recrystallized, or so tightly cemented together, that the rock breaks across the grains rather than around them. See p. 8.

Quaternary: The most recent geologic period in the Cenozoic. It includes the Pleistocene Epoch and the Recent. See Table IV on p. 17.

Range: A high-standing, geographically distinct, assemblage of mountains. In our area, this term is also applied to the resistant, linear ridges of Precambrian rock exposed in the historic iron and copper mining districts of northern Minnesota, northern Wisconsin, and upper Michigan. See p. 30 and p. 88.

Recent: The most recent epoch in the Quaternary geologic period; includes the present. See Table IV on p. 17.

Recessional moraine: Till deposited in ridges at the terminus of a retreating glacier when temporary pauses occurred. See end moraine above and p. 54.

Reef: A massive, wave-resistant, moundlike structure formed by carbonate-secreting colonies of plants and animals. In our area, such features are present in Silurian rocks. See p. 48 and p. 169.

Reef core: The massive, central part of a reef complex that was bound into a rigid mass by colonial forms of carbonate-secreting plants and animals. See p. 170 and Fig. 6-18 on p. 171.

Reef flank: The thick-bedded part of a reef complex that dips radially away from the reef core. The material composing these beds represents debris removed from the reef by wave action. See reef core above and Fig. 6-18 on p. 171.

Regression: Retreat of the sea from the land. This can be accomplished by a worldwide lowering of sea level, or by the uplift of a land area. See p. 18 and Fig. 1-12 on p. 19.

Reverse fault: A break in a rock sequence where the displacement is such that movement was up for rocks above the inclined fault plane relative to those below it. See Fig. 1-14 on p. 22.

Rhyolite: A fine-grained, generally extrusive, acidic igneous rock. In our area, rhyolites are often reddish in color and porphyritic. See p. 1 and Table I on p. 7.

Ripple marks: A gently undulatory surface consisting of alternating, subparallel, small-scale ridges and troughs developed in sand or silt by wind, wave, or current action. Such features are often preserved in rocks. See p. 120.

Rock cycle: The concept that interrelates the major rock types (igneous, metamorphic, and sedimentary) into an idealized cycle of events that includes crystallization of igneous rocks from magma, uplift and erosion of these rocks, deposition and lithification of the eroded material to form sedimentary rocks, increasing accumulation of sediments resulting in deep burial, metamorphism by heat and pressure, and a continuation of this process so that fusion produces another magma to initiate the sequence again. See p. 1 and Fig. 1-1 on p. 2.

Rock flour: Extremely fine, rock particles formed by the grinding of debris in glacial ice against the underlying bedrock. See p. 54.

Rock unit: A sequence of rocks, or a rock body, that can be readily distinguished from adjacent rock units. See formation above.

Sandstone: A clastic sedimentary rock primarily composed of sand-size mineral grains or rock particles that are compacted and/or cemented together.

Schist: A strongly foliated, medium-grade, metamorphic rock that commonly contains muscovite and quartz, in addition to other minerals. See p. 4 and Table III on p. 9.

Sedimentary rock: A rock formed by the lithification of: particles of various sizes that were derived from the erosion of preexisting rocks, chemical precipitates, and/or organic matter of plant or animal origin. These materials usually were deposited in layers or beds. See p. 3.

Septa: Internal partitions that divide the shell or exoskeleton of many animals.

Shale: A sedimentary rock composed of silt and clay that parts along closely spaced planes essentially parallel to bedding. See Table II on p. 8.

Shield: A large area of Precambrian bedrock, usually located near the center of a continent. In our area, this is the Canadian shield. See p. 23 and Fig. 2-1 on p. 24.

Siliceous: An adjective used to describe a rock that contains abundant silica.

Sill: A tabular, intrusive, igneous rock body that is concordant with surrounding rock layers. See p. 1 and Fig. 1-2 on p. 3.

Sillimanite: A glassy to whitish, needlelike, aluminum silicate mineral that forms in high-rank metamorphic rocks. See p. 9.

Silt: A mineral grain or rock fragment within the size range from 1/256 mm to 1/16 mm; smaller than a fine sand grain and larger than a clay particle.

Siltstone: A clastic sedimentary rock primarily composed of mineral grains or rock fragments of silt-size. See silt above and Table II on p. 8.

Silurian: The third geologic period of the Paleozoic Era. The Silurian is sometimes referred to as the age of corals since their remains are especially common in rocks of this age. See Table IV on p. 17.

Sinkhole: A generally circular depression of the land surface, which is created by the collapse of a cave. See cave above and and Fig. 5-2 on p. 113.

Slate: A fine-grained, low-rank, metamorphic rock which exhibits a well-developed cleavage (plane of parting) that is at some angle to the bedding. See p. 4 and Table III on p. 9.

Slump block: A mass of material on a hill, cliff, or bluff that rotates downward and outward during slippage along an arcuate plane of failure. See Fig. 6-9 on p. 159.

Soil profile: A vertical section through a soil into the sediment or bedrock below. Such a sequence displays the various soil horizons, and the relationship of soil development to the underlying material. See Fig. 8-5 on p. 196.

Sphalerite: A yellowish to dark brown, resinous-appearing, zinc sulfide mineral that displays six well-developed cleavages. Sphalerite is a common ore of zinc in the Upper Mississippi Valley Lead-Zinc District. See p. 191.

Spit: A narrow projection of sand, deposited by longshore currents, that extends into a bay, or outward from a headland. See Fig. 1-9 on p. 13.

Stalactite: A calcite (calcium carbonate) deposit resembling an icicle that hangs from the roof of a cave. Such features are deposited by the evaporation of dripping water that contains calcium bicarbonate in solution. See Fig. 5-16 on p. 130.

Stalagmite: An elongate, conical deposit of calcite (calcium carbonate) that develops upward from the floor of a cave due to the evaporation of dripping water that contains calcium bicarbonate in solution. See Fig. 5-16 on p. 130.

Staurolite: A brown to black, prismatic, iron, magnesium, aluminum silicate mineral which often forms as a crystal intergrowth (called a twin) that resembles a small cross. This mineral is usually found in medium-rank metamorphic rocks such as schist.

Strata: Distinct layers or beds in sedimentary and extrusive igneous rocks.

Stratigraphic: An adjective that refers to the study and/or interpretation of sequences of layered or bedded rocks.

Stratigraphic column: A diagram drawn in columnar fashion to show the relative age and position of a sequence of rocks in a given area. In such illustrations the oldest rocks are at the bottom and the youngest are at the top. It is also used to refer to the rock succession in a given area.

Striation, Striae: A scratch(es) or groove(s) inscribed on a rock surface by a moving geologic agent such as a glacier, a stream, or by faulting. See p. 53 and Fig. 2-39 on p. 65.

Strike: The trend of a line defined by the intersection of an imaginary horizontal plane with an inclined rock surface. See p. 21.

Subsidence: The downwarping or sinking of a part of the earth's crust. See Fig. 1-3 on p. 4.

Superior Upland: The southern part of the Canadian shield that projects southward into the United States. The Northern Highland of our region is included within this province. See Fig. 2-2 on p. 25.

Superposition, Law of: The principle which states that previously deposited materials (older rocks) in an undisturbed sequence underlie more recently deposited materials (younger rocks). See p. 14 and Fig. 1-10 on p. 15.

Syncline: A troughlike downfold where the youngest rocks are in the center of the structure. See Fig. 1-14 on p. 22.

Taconite: A banded, often reddish, rock composed of silica-rich and iron-rich layers. Such rocks are currently utilized as an ore of iron. They ordinarily contain only about 25% iron, and they must be enriched by artificial concentration processes. See iron formation above, p. 30, and p. 89.

Talus: An accumulation of angular rock debris at the foot of a bedrock slope or cliff. See p. 124.

Tectonic: An adjective that generally refers to structural features in rocks that were produced by deformational forces, or to the historical development of such features in a given region.

Terminal moraine: A till ridge (or ridges) that formed along the ice front during the maximum advance of a glacier. See end moraine above and p. 54.

Tertiary: The oldest geologic period in the Cenozoic Era. See Table IV on p. 17.

Texture: The size, shape, and arrangement of all the constituents that make up a rock. This includes the distribution, orientation, and mutual relations between the various particles, grains, crystals, cements, etc. See Fig. 1-4 on p. 6.

Thorax: The segmented, central body portion of a trilobite. See p. 180 and Fig. 7-5 on p. 181.

Thrust fault: A type of reverse fault in which the fault plane is inclined at a low angle (generally 25 degrees or less). See reverse fault above and Fig. 1-14 on p. 22.

Tidal zone: The coastal area that lies between low and high tide. Consequently, this zone is alternately covered and uncovered by tidal fluctuations.

Till: Unsorted, unstratified glacial drift which was deposited directly by the ice. See p. 53 and Fig. 2-27 on p. 54.

Tongue: A term we use to describe a relatively narrow lobe of glacial ice that projected outward from the main ice sheet.

Topography: Configuration of the land surface in a given area.

Transgression: The advance of the sea onto the land. This can be accomplished by a worldwide rise in sea level, or by the depression of a land area. See p. 18 and Fig. 1-12 on p. 19.

Triassic: The first geologic period in the Mesozoic Era. See Table IV on p. 17.

Trilobite: An extinct marine arthropod that was a distant relative of the present-day horseshoe crab. See p. 180 and Fig. 7-5 on p. 181.

Two Creekan: The fourth substage of the Wisconsinan glacial stage. It represents an interval of moderating climate, and a retreat of the glacial ice sheet between the Woodfordian and Valderan advances. See Table VI on p. 63.

Type section: A distinctive rock unit in a given area that was rigorously defined and named for a nearby geographic feature. As such, it becomes a standard to which other, possibly similar, rock units may be compared. The formal name at the type section is extended outward from the type locality as far as the distinctive characteristics can be recognized. See p. 19.

Unconformity: The general term for an interruption in the geologic record due to erosion or nondeposition. See p. 14.

Uniformitarianism, Principle of: The doctrine which may be stated as: "The present is the key to the past." Based on this principle, the geologic processes that are operating on the surface of the earth today are like those that operated in the past. See p. 3.

Uplift: An upwarp of the earth's crust. See Fig. 1-3 on p. 4.

Valderan: The substage of the Wisconsinan glacial stage which encompasses the latest glacial advance. See p. 62 and Table VI on p. 63.

Vesicles: Voids in lava flows that formed when gas bubbles were entrapped in the molten rock as it crystallized. See p. 1 and p. 2.

Vitreous: An adjective that means glasslike, or describes a luster like that of glass.

Volcano: A generally conical build-up of lava and/or volcanic ejecta around a vent at the surface of the earth through which magma, ejecta, and gases reach the surface. See Fig. 1-2 on p. 3.

Vug: A small solution cavity in a carbonate rock.

Water gap: A narrow pass occupied by a stream that cuts through a bedrock ridge. Such features formed when streams that were well-established on a surface above a buried ridge continued downcutting into the resistant rocks below. See p. 90.

Water table: See ground water table above.

Western Uplands: The generally unglaciated, highland area in the western part of our region that is highly dissected by stream erosion, and characterized by a series of resistant, cuesta-forming rock units that alternate with less resistant rocks. See p. 72, Fig. 3-1 on p. 73, and Chapter 5.

Wind-frosted: A lusterless, very finely-pitted, surface that resembles etched or ground glass. Such a surface is produced when the wind blows sand grains against each other, or against a rock surface.

Wind gap: A former water gap that was abandoned when the erosive power of the stream was no longer able to down-cut through the buried, resistant, bedrock ridge, or when the stream course was diverted by some other mechanism. See water gap above, p. 85, and p. 90.

Woodfordian: The third substage of the Wisconsinan glacial stage. This substage represents a major advance of Late Wisconsinan glacial ice. See p. 62 and Table VI on p. 63.

INDEX

Note: Most geographical features are identified as to state by (IL) for Illinois, (IO) for Iowa, (MI) for Michigan, (MN) for Minnesota, and (WI) for Wisconsin.